Methods in Molecular Biology

Series Editor
John M. Walker
School of Life and Medical Sciences
University of Hertfordshire
Hatfield, Hertfordshire, AL10 9AB, UK

For further volumes:
http://www.springer.com/series/7651

Programmed Necrosis

Methods and Protocols

Edited by

Adrian T. Ting

Precision Immunology Institute, Icahn School of Medicine at Mount Sinai, New York, NY, USA

Editor
Adrian T. Ting
Precision Immunology Institute
Icahn School of Medicine at Mount Sinai
New York, NY, USA

ISSN 1064-3745 ISSN 1940-6029 (electronic)
Methods in Molecular Biology
ISBN 978-1-4939-8753-5 ISBN 978-1-4939-8754-2 (eBook)
https://doi.org/10.1007/978-1-4939-8754-2

Library of Congress Control Number: 2018950849

Cover illustration: Printed with permission from © Mount Sinai Health System. Ni-ka Ford © Mount Sinai Health System

This Humana Press imprint is published by the registered company Springer Science+Business Media, LLC part of Springer Nature.
The registered company address is: 233 Spring Street, New York, NY 10013, U.S.A.

Preface

Cells undergo many different forms of regulated cell death including apoptosis and programmed necrosis. The latter death program includes necroptosis, which is a lytic form of cell death. Necroptosis can be triggered by members of the TNF family and some pattern recognition receptors. It occurs in the absence of caspase activity and is defined as dependent on the kinase RIPK3 and its downstream effector molecule MLKL. At the cellular level, it is characterized by the loss of membrane integrity and the release of cellular contents that function as damage-associated molecular patterns (DAMPs) or alarmins. Thus, necroptosis is considered to be an inflammatory and immunogenic form of cell death.

The aim of this volume is to provide readers with the background on this emerging form of cell death, and the techniques and approaches that are used to study it. Since the role of necroptosis in normal physiology and in diseases is not well understood, we hope that this volume will be a useful resource for research technicians, graduate students, postdoctoral fellows, and other scientists who are pursuing these questions.

New York, NY, USA *Adrian T. Ting*

Contents

Preface . . . . *v*
Contributors . . . . *ix*

1 Tools in the Art of Studying Necroptosis 1
Adrian T. Ting

2 Loss-of-Function RNAi Screen to Identify Necrosis-Signaling Molecules 11
David Mark Moquin and Francis Ka-Ming Chan

3 Chemical Library Screens to Identify Pharmacological Modulators of Necroptosis 19
Danish Saleh and Alexei Degterev

4 Distinguishing Necroptosis from Apoptosis 35
Inbar Shlomovitz, Sefi Zargarian, Ziv Erlich, Liat Edry-Botzer, and Motti Gerlic

5 Methods for Studying TNF-Mediated Necroptosis in Cultured Cells 53
Zikou Liu, John Silke, and Joanne M. Hildebrand

6 Analysis of Necroptosis in Bone Marrow-Derived Macrophages 63
Diana Legarda and Adrian T. Ting

7 Generation and Use of Chimeric RIP Kinase Molecules to Study Necroptosis 71
Diego A. Rodriguez and Douglas R. Green

8 Detection of MLKL Oligomerization During Programmed Necrosis 85
Zhenyu Cai and Zheng-Gang Liu

9 Analysis of Cytokine- and Influenza A Virus-Driven RIPK3 Necrosome Formation. 93
Roshan J. Thapa, Shoko Nogusa, and Siddharth Balachandran

10 Detection of RIPK1 in the FADD-Containing Death Inducing Signaling Complex (DISC) During Necroptosis. 101
Rosalind L. Ang and Adrian T. Ting

11 Use of RIP1 Kinase Small-Molecule Inhibitors in Studying Necroptosis 109
Allison M. Beal, John Bertin, and Michael A. Reilly

12 Analyzing Necroptosis Using an RIPK1 Kinase Inactive Mouse Model of TNF Shock 125
Matija Zelic and Michelle A. Kelliher

13 Assessment of In Vivo Kidney Cell Death: Acute Kidney Injury 135
Wulf Tonnus, Moath Al-Mekhlafi, Christian Hugo, and Andreas Linkermann

14 Assessment of In Vivo Kidney Cell Death: Glomerular Injury 145
Wulf Tonnus, Moath Al-Mekhlafi, Florian Gembardt, Christian Hugo, and Andreas Linkermann

15 Detection of Necroptosis by Phospho-RIPK3 Immunohistochemical Labeling 153
Joshua D. Webster, Margaret Solon, Susan Haller, and Kim Newton

16 Characterization of the TNFR1-SC Using "Modified Tandem Affinity Purification" in Conjunction with Liquid Chromatography–Mass Spectrometry (LC-MS) 161
Matthias Reichert, Amandeep Bhamra, Sebastian Kupka, and Henning Walczak

17 Monitoring RIPK1 Phosphorylation in the TNFR1 Signaling Complex 171
Dario Priem, Yves Dondelinger, and Mathieu J. M. Bertrand

18 Analysis of CYLD Proteolysis by CASPASE 8 in Bone Marrow-Derived Macrophages 181
Diana Legarda and Adrian T. Ting

Index *189*

Contributors

MOATH AL-MEKHLAFI • *Division of Nephrology, Department of Internal Medicine III, University Hospital Carl Gustav Carus, Technische Universität Dresden, Dresden, Germany*

ROSALIND L. ANG • *Precision Immunology Institute, Icahn School of Medicine at Mount Sinai, New York, NY, USA*

SIDDHARTH BALACHANDRAN • *Blood Cell Development and Function Program, Fox Chase Cancer Center, Philadelphia, PA, USA*

ALLISON M. BEAL • *Pattern Recognition Receptor Discovery Performance Unit, Immuno-Inflammation Therapeutic Area, GlaxoSmithKline, Collegeville, PA, USA*

JOHN BERTIN • *Pattern Recognition Receptor Discovery Performance Unit, Immuno-Inflammation Therapeutic Area, GlaxoSmithKline, Collegeville, PA, USA*

MATHIEU J. M. BERTRAND • *VIB Center for Inflammation Research, Zwijnaarde-Ghent, Belgium; Department of Biomedical Molecular Biology, Ghent University, Zwijnaarde-Ghent, Belgium*

AMANDEEP BHAMRA • *Proteomics Research Core Facility, UCL Cancer Institute, University College London, London, UK*

ZHENYU CAI • *Center for Cancer Research, National Cancer Institute, National Institutes of Health, Bethesda, MD, USA; National Center for Liver Cancer, Shanghai, China*

FRANCIS KA-MING CHAN • *Department of Immunology, Duke University Medical Center, Durham, NC, USA*

ALEXEI DEGTEREV • *Department of Developmental, Molecular and Chemical Biology, Tufts University School of Medicine, Boston, MA, USA*

YVES DONDELINGER • *VIB Center for Inflammation Research, Zwijnaarde-Ghent, Belgium; Department of Biomedical Molecular Biology, Ghent University, Zwijnaarde-Ghent, Belgium*

LIAT EDRY-BOTZER • *Department of Clinical Microbiology and Immunology, Sackler Faculty of Medicine, Tel Aviv University, Tel Aviv, Israel*

ZIV ERLICH • *Department of Clinical Microbiology and Immunology, Sackler Faculty of Medicine, Tel Aviv University, Tel Aviv, Israel*

FLORIAN GEMBARDT • *Division of Nephrology, Department of Internal Medicine III, University Hospital Carl Gustav Carus, Technische Universität Dresden, Dresden, Germany*

MOTTI GERLIC • *Department of Clinical Microbiology and Immunology, Sackler Faculty of Medicine, Tel Aviv University, Tel Aviv, Israel*

DOUGLAS R. GREEN • *Department of Immunology, St. Jude Children's Research Hospital, Memphis, TN, USA*

SUSAN HALLER • *Department of Pathology, Genentech, South San Francisco, CA, USA*

JOANNE M. HILDEBRAND • *Walter and Eliza Hall Institute of Medical Research, Parkville, VIC, Australia; Department of Medical Biology, University of Melbourne, Parkville, VIC, Australia*

CHRISTIAN HUGO • *Division of Nephrology, Department of Internal Medicine III, University Hospital Carl Gustav Carus, Technische Universität Dresden, Dresden, Germany*

MICHELLE A. KELLIHER • *Department of Molecular, Cell and Cancer Biology, University of Massachusetts Medical School, Worcester, MA, USA*

SEBASTIAN KUPKA • *Centre for Cell Death, Cancer and Inflammation (CCCI), UCL Cancer Institute, University College London, London, UK*

DIANA LEGARDA • *Precision Immunology Institute, Icahn School of Medicine at Mount Sinai, New York, NY, USA*

ANDREAS LINKERMANN • *Division of Nephrology, Department of Internal Medicine III, University Hospital Carl Gustav Carus, Technische Universität Dresden, Dresden, Germany*

ZHENG-GANG LIU • *Center for Cancer Research, National Cancer Institute, National Institutes of Health, Bethesda, MD, USA*

ZIKOU LIU • *Walter and Eliza Hall Institute of Medical Research, Parkville, VIC, Australia; Department of Medical Biology, University of Melbourne, Parkville, VIC, Australia*

DAVID MARK MOQUIN • *Department of Immunology, Duke University Medical Center, Durham, NC, USA*

KIM NEWTON • *Department of Physiological Chemistry, Genentech, South San Francisco, CA, USA*

SHOKO NOGUSA • *Blood Cell Development and Function Program, Fox Chase Cancer Center, Philadelphia, PA, USA*

DARIO PRIEM • *VIB Center for Inflammation Research, Zwijnaarde-Ghent, Belgium; Department of Biomedical Molecular Biology, Ghent University, Zwijnaarde-Ghent, Belgium*

MATTHIAS REICHERT • *Centre for Cell Death, Cancer and Inflammation (CCCI), UCL Cancer Institute, University College London, London, UK*

MICHAEL A. REILLY • *Pattern Recognition Receptor Discovery Performance Unit, Immuno-Inflammation Therapeutic Area, GlaxoSmithKline, Collegeville, PA, USA*

DIEGO A. RODRIGUEZ • *Department of Immunology, St. Jude Children's Research Hospital, Memphis, TN, USA*

DANISH SALEH • *Medical Scientist Training Program, Program in Neuroscience, Sackler School of Graduate Biomedical Sciences, Tufts University School of Medicine, Boston, MA, USA*

INBAR SHLOMOVITZ • *Department of Clinical Microbiology and Immunology, Sackler Faculty of Medicine, Tel Aviv University, Tel Aviv, Israel*

JOHN SILKE • *Walter and Eliza Hall Institute of Medical Research, Parkville, VIC, Australia; Department of Medical Biology, University of Melbourne, Parkville, VIC, Australia*

MARGARET SOLON • *Department of Pathology, Genentech, South San Francisco, CA, USA*

ROSHAN J. THAPA • *Blood Cell Development and Function Program, Fox Chase Cancer Center, Philadelphia, PA, USA*

ADRIAN T. TING • *Precision Immunology Institute, Icahn School of Medicine at Mount Sinai, New York, NY, USA*

WULF TONNUS • *Division of Nephrology, Department of Internal Medicine III, University Hospital Carl Gustav Carus, Technische Universität Dresden, Dresden, Germany*

HENNING WALCZAK • *Centre for Cell Death, Cancer and Inflammation (CCCI), UCL Cancer Institute, University College London, London, UK*
JOSHUA D. WEBSTER • *Department of Pathology, Genentech, South San Francisco, CA, USA*
SEFI ZARGARIAN • *Department of Clinical Microbiology and Immunology, Sackler Faculty of Medicine, Tel Aviv University, Tel Aviv, Israel*
MATIJA ZELIC • *Department of Molecular, Cell and Cancer Biology, University of Massachusetts Medical School, Worcester, MA, USA*

Chapter 1

Tools in the Art of Studying Necroptosis

Adrian T. Ting

Abstract

Necroptosis is a more recently described form of regulated cell death (RCD) that occurs in a caspase-independent manner. This is a lytic form of cell death in which the cellular contents are released and these contents serve as damage-associated molecular patterns (DAMPs). DAMPs are endogenous ligands for pattern recognition receptors and therefore necroptosis is considered to be highly inflammatory and immunogenic. Members of the TNF family are the most well-studied triggers of necroptosis, though other immune receptors are also known to directly trigger this death pathway. Necroptosis is now defined to be dependent on the core signaling molecules of RIPK3 and MLKL. In the case of TNF, it also involves RIPK1, and induction of necroptosis often also requires inhibition of caspases. In this volume, a wide array of tools to study necroptosis are described. They include pharmacological, biochemical, cellular, and in vivo approaches. Combining multiple approaches in one's study is ideal for generating conclusive evidence for the involvement of necroptosis. The protocols presented in this chapter are highly useful in studying necroptosis, whose physiological and pathophysiological roles remain incompletely understood.

Key words Regulated cell death, Apoptosis, Necroptosis, Programmed necrosis, Tumor necrosis factor (TNF), Receptor-interacting protein kinase (RIPK), Mixed lineage kinase domain-like (MLKL)

1 Programmed Necrosis

Regulated cell death (RCD) is a cellular process that underlies a number of physiological and pathophysiological responses. The field has been dominated by the study of apoptosis, a term first introduced to describe a form of cellular demise observed in ischemic liver cells [1], characterized by morphological features such as chromatin condensation and plasma membrane blebbling. Cellular contents from such dying cells are contained within these apoptotic bodies, which are taken up by neighboring phagocytes. Apoptosis is regulated by CASPASE 3, 7, 8, 9, and 10 and members of the BCL-2 family [2], and the role of this form of RCD in a number of biological processes and human diseases is now well appreciated.

Other forms of RCD with morphological features of cellular swelling and rupture, distinct from that of apoptosis, have also

Adrian T. Ting (ed.), *Programmed Necrosis: Methods and Protocols*, Methods in Molecular Biology, vol. 1857, https://doi.org/10.1007/978-1-4939-8754-2_1, © Springer Science+Business Media, LLC, part of Springer Nature 2018

been observed and these alternative death programs have garnered significant interest in recent years. An understanding of the biological functions of these nonapoptotic death programs is only starting to emerge with the recent identification of distinct molecules that regulate these death programs. The first of these lytic RCD to be characterized was termed pyroptosis, because it was associated with the release of pyrogenic cytokines IL-1β and IL-18, as well as damage-associated molecular pattern (DAMPs) to generate a proinflammatory response [3, 4]. CASPASE 1-mediated cleavage of precursor polypeptides generates mature IL-1β and IL-18, whereas CASPASE 4/5/11-mediated cleavage of gasdermin D causes membrane pore formation and cell lysis [5]. Another lytic RCD that has been described is necroptosis, and this term was coined to describe nonapoptotic cell death that was occurring in response to ligation of death receptors (DR) of the TNF receptor superfamily (TNFRSF) in the presence of caspase inhibitors [6]. Cells undergoing this form of RCD possess features of necrosis and have also been called programmed necrosis [7]. Since the term programmed necrosis could be used to describe pyroptosis and other forms of necrosis [5], the more widely used term is now necroptosis, which is defined as RCD dependent on the core signaling molecules of RIPK3 and MLKL [8]. Since the techniques used to study pyroptosis have been described in other volumes in this series, this volume will primarily describe techniques that are used to analyze necroptosis.

2 Necroptosis Signaling Molecules

The initial signaling molecule discovered to be necessary for DR-induced necroptosis was RIPK1 and importantly, the kinase domain of RIPK1 was shown to be essential for this process [9]. The subsequent identification of the compound necrostatin-1 as an inhibitor of necroptosis via inhibiting the kinase activity of RIPK1 [6, 10] further implicated RIPK1 in regulating necroptosis. The seminal discovery of RIPK3 as an essential molecule in this pathway, through a homotypic RHIM-dependent interaction with RIPK1, provided a critical marker for defining necroptosis [11–13]. RIPK3 phosphorylates the pseudokinase MLKL [14, 15], which undergoes oligomerization and translocates to the plasma membrane. The precise mechanism by which oligomeric MLKL triggers cell lysis remains unclear and could occur through the direct formation of transmembrane pores or by activating ion channels to effect ion influx [16–19]. While TNF and its receptors are the most well studied inducer of necroptosis, other receptors are also capable of inducing necroptosis directly (Fig. 1). In

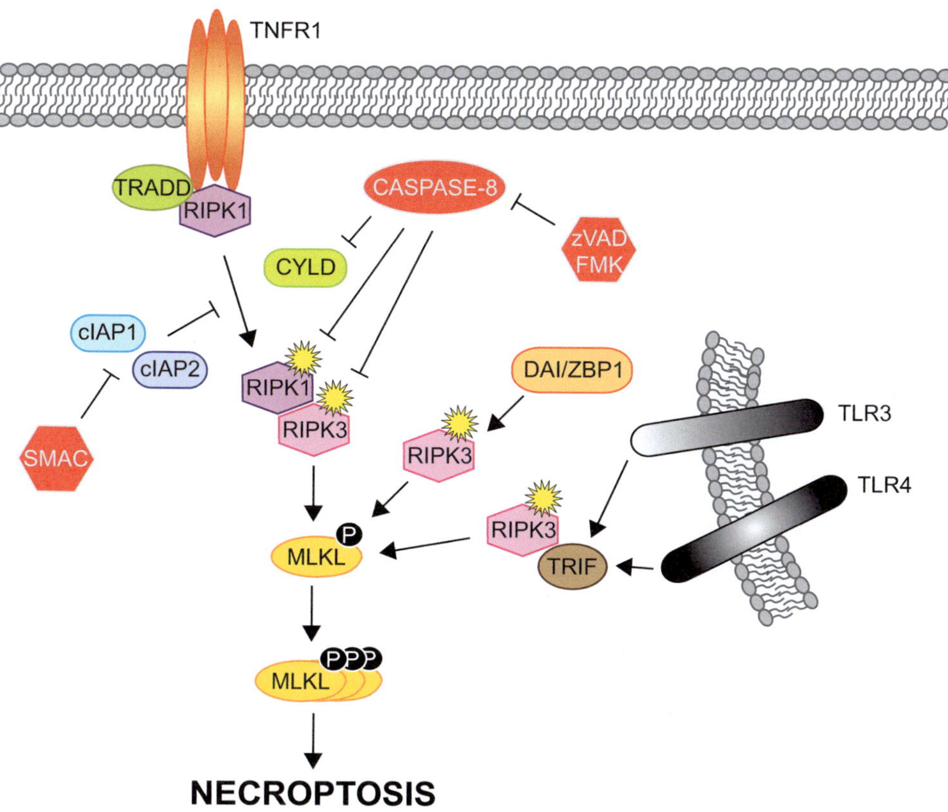

Fig. 1 The necroptosis pathway has been most studied following stimulation of cells with TNF. TNFR1 utilizes RIPK1 to propagate the death signal to RIPK3, which then phosphorylates MLKL. Phosphorylated MLKL forms oligomers that cause cell lysis. In many cases, experimental induction of necroptosis by TNF requires (1) conversion of RIPK1 to become a death signaling molecule by blocking its ubiquitination, usually by inhibiting its E3 ligases cIAP1/2 using SMAC mimetics; and (2) inhibition of CASPASE 8 using a caspase inhibitor such as zVAD-FMK because CASPASE 8 antagonizes the necroptosis pathway. Other receptors such as the pattern recognition receptors DAI/ZBP1, TLR3, and TLR4 can also activate necroptosis. DAI/ZBP1 can directly interact with RIPK3, whereas TLR3 and TLR4 utilizes the TRIF adapter to signal to RIPK3

generally, these receptors either encode a RHIM domain (e.g., ZBP1/DAI) or utilize the RHIM-containing signaling adapter TRIF (e.g., TLR3 and TLR4). These RHIM-containing molecules interact directly with the RHIM on RIPK3, bypassing the need for RIPK1, culminating in the phosphorylation and oligomerization of MLKL. Studying necroptosis often requires the genetic deletion of CASPASE 8 in vivo or its pharmacological inhibition using caspase inhibitors in vitro. CASPASE 8 inhibits necroptosis by proteolytic cleavage of the necroptosis signaling molecules RIPK1, RIPK3, and CYLD [20–22]. Apoptosis is thought to be the default death response in many different cell types, at least in response to DR ligation. The current paradigm is that necroptosis provides a “trap door” during infections to overcome microbial blockade of caspases and apoptosis by switching to necroptotic death [23]. This would deny microbes a host cell for replication, as well as generate an inflammatory response via the release of DAMPs. It is also possible that necroptosis serves to attenuate inflammation by killing off inflammatory cells that are producing cytokines, as has been recently proposed [24].

3 The Necroptosis Toolkit

The study of necroptosis requires numerous pharmacological, biochemical, molecular, and in vivo techniques that will be described in the following chapters. A combination of different techniques is often needed to conclusively demonstrate that necroptosis is involved in the particular system one is analyzing. Moquin and Chan describe a loss-of-function screen using RNA interference to discover genes that regulate necroptosis, whereas Saleh and Degterev describe a strategy to screen a chemical library to identify novel compounds that affect this death pathway. Since apoptosis and necroptosis are related RCD pathways, techniques to distinguish between the two death modalities are critically important. Historically, apoptotic cells were defined as staining positive by annexin V but negative for propidium iodide (or other membrane impermeant dyes), whereas necrotic cells were defined as staining negative by annexin V but positive for propidium iodide. This definition was formulated based on staining with necrotic cells that were lysed by freeze-thawing. Freeze-thawed cells differ significantly from cells undergoing necroptosis, which do stain positive with annexin V [22, 25]. Approaches to distinguish between apoptosis and necroptosis are now described by Shlomovitz et al. in Chapter 4. Methods to analyze necroptosis in primary fibroblast cultures are described by Liu et al. in Chapter 5 and in primary macrophage cultures by Legarda and Ting in Chapter 6. In Chapter 7, Rodriguez and Green describe an approach using chimeric RIP kinase molecules, which is particularly useful in studying the more distal necroptosis pathway subsequent to the activation of RIP kinases [25, 26]. Currently, the most definitive biochemical assay to detect necroptosis is to examine MLKL oligomerization, and this assay is described by Cai and Liu in Chapter 8. Two additional complementary biochemical assays based on coimmunoprecipitation are also provided here. In Chapter 9, Thapa et al. detail how the RIPK3-containing necrosome complex can be studied while in Chapter 10, Ang and Ting describe how RIPK1 association with FADD during necroptosis can be detected.

Pharmacological tools have been invaluable in studying necroptosis as exemplified by the discovery and use of necrostatin-1 to block RIPK1. In Chapter 11, Beal et al. provide an in-depth discussion of additional RIPK1 inhibitors and their use in both in vitro and in vivo settings to study the role of RIPK1 in cell death and inflammation. Pharmacological approaches can be complemented by genetic approaches and in Chapter 12, Zelic and Kelliher describe the use of the *Ripk1*^{D138N} kinase-inactive mouse strain. Tonnus et al. provide several additional in vivo mouse models in Chapters 13 and 14, where cell death and kidney injury induced by surgical procedures or drugs can be analyzed. One difficulty in studying necroptosis has been the in situ documentation of this RCD occurring in tissues during physiological responses and in disease states. One strategy has been to detect MLKL phosphorylation, which is necessary for its oligomerization, using a phospho-MLKL antibody. This has been described for human MLKL phosphorylated on Thr^{357} and Ser^{358} with human tissues [18] while in vivo mouse studies have been hampered by the lack of a similar reagent. The recent description of a monoclonal antibody that detects phosphorylation on Thr^{231} and Ser^{232} of murine RIPK3 [27], which occurs during necroptosis, will now enable mouse studies to be conducted. Webster et al. describe the use of this monoclonal antibody in immunohistochemistry of tissue sections in Chapter 15.

It should be noted that cell death is often prevented from occurring due to antagonism by cell survival mechanisms. A good example is provided by the response to TNF or other ligands in the TNF family where the default cellular response is survival. One key survival mechanism in the TNF pathway is the nondegradative ubiquitination and phosphorylation of RIPK1 [28]. In order to sensitize cells to TNF-induced necroptosis, ubiquitination of RIPK1 has to be disrupted. This is often achieved using pharmacological mimics of SMAC to antagonize cIAP1/2, the E3 ligases that ubiquitinate RIPK1 (Fig. 1), in addition to caspase inhibitors. Thus, in the context of studying necroptosis, it is often instructive to examine whether these survival mechanisms are being modulated. In Chapter 16, Reichert et al. describe a biochemical approach to purify and characterize the TNF receptor complex. Perturbation in this signaling complex often alters sensitivity to cell death including necroptosis [29]. In Chapter 17, Priem et al. provide a biochemical protocol to examine RIPK1 phosphorylation, another key survival signal that antagonizes necroptosis [30]. Finally in Chapter 18, Legarda and Ting describe a protocol to analyze the proteolytic cleavage of the deubiquitinase CYLD, a prosurvival signaling event that inhibits TNF-induced necroptosis [31]. Methodologies that analyze these prosurvival mechanisms provide a counterpoint to those that analyze necroptosis, and therefore examining both sides of the coin provides a more comprehensive picture.

4 Considerations in the Use of Different Tools

As the field evolved and we gain more understanding of necroptosis, the limitations of strategies that are used to study this form of RCD are also becoming better understood. For instance, blocking the kinase activity of RIPK1 using necrostatin-1 has been widely used but this compound can also inhibit IDO [32] and this caveat should be kept in mind. Furthermore, the kinase activity of RIPK1 also participates in RIPK1-dependent apoptosis, as well as in nondeath functions [33, 34]. Therefore, the involvement of the RIPK1 kinase, in and of itself, is insufficient to demonstrate necroptosis. RIPK3 knockout mice and cells have also been used extensively to study necroptosis. It is now clear that RIPK3 also participates in non-necroptotic functions [34–36]. A comparative study has shown RIPK1 and RIPK3 mutant mice to behave differently from MLKL mutant mice in a number of experimental mouse models of inflammation [37] indicating that RIPK1 and RIPK3 have functions beyond necroptosis. Therefore, observations obtained from RIPK1 or RIPK3 mutant mice may not be definitive for necroptosis whereas those from MLKL mutant mice provide more conclusive evidence for the involvement of necroptosis. These are points to consider when choosing the appropriate approaches to use to study necroptosis, and combining several of them is strongly encouraged to overcome limitations inherent to individual approaches. As the regulation of this cell death pathway and its biological functions remain to be fully understood, the methodologies described in this volume should provide investigators with a variety of tools to address these open questions.

Acknowledgments

Studies from the author's laboratory were supported by grants AI052417, AI126036, and AI132405 from the National Institutes of Health. I thank Dr. Diana Legarda for generating Fig. 1 and for critically reading the manuscript.

References

1. Kerr JF, Wyllie AH, Currie AR (1972) Apoptosis: a basic biological phenomenon with wide-ranging implications in tissue kinetics. Br J Cancer 26(4):239–257
2. Green DR, Llambi F (2015) Cell death signaling. Cold Spring Harb Perspect Biol 7(12):a006080. https://doi.org/10.1101/cshperspect.a006080

3. Cookson BT, Brennan MA (2001) Pro-inflammatory programmed cell death. Trends Microbiol 9(3):113–114
4. Fink SL, Cookson BT (2005) Apoptosis, pyroptosis, and necrosis: mechanistic description of dead and dying eukaryotic cells. Infect Immun 73(4):1907–1916. https://doi.org/10.1128/IAI.73.4.1907-1916.2005
5. Wallach D, Kang TB, Dillon CP, Green DR (2016) Programmed necrosis in inflammation: toward identification of the effector molecules. Science 352(6281):aaf2154. https://doi.org/10.1126/science.aaf2154
6. Degterev A, Huang Z, Boyce M, Li Y, Jagtap P, Mizushima N, Cuny GD, Mitchison TJ, Moskowitz MA, Yuan J (2005) Chemical inhibitor of nonapoptotic cell death with therapeutic potential for ischemic brain injury. Nat Chem Biol 1(2):112–119. https://doi.org/10.1038/nchembio711
7. Chan FK, Shisler J, Bixby JG, Felices M, Zheng L, Appel M, Orenstein J, Moss B, Lenardo MJ (2003) A role for tumor necrosis factor receptor-2 and receptor-interacting protein in programmed necrosis and antiviral responses. J Biol Chem 278(51):51613–51621
8. Galluzzi L, Vitale I, Aaronson SA, Abrams JM, Adam D, Agostinis P, Alnemri ES, Altucci L, Amelio I, Andrews DW, Annicchiarico-Petruzzelli M, Antonov AV, Arama E, Baehrecke EH, Barlev NA, Bazan NG, Bernassola F, Bertrand MJM, Bianchi K, Blagosklonny MV, Blomgren K, Borner C, Boya P, Brenner C, Campanella M, Candi E, Carmona-Gutierrez D, Cecconi F, Chan FK, Chandel NS, Cheng EH, Chipuk JE, Cidlowski JA, Ciechanover A, Cohen GM, Conrad M, Cubillos-Ruiz JR, Czabotar PE, D'Angiolella V, Dawson TM, Dawson VL, De Laurenzi V, De Maria R, Debatin KM, DeBerardinis RJ, Deshmukh M, Di Daniele N, Di Virgilio F, Dixit VM, Dixon SJ, Duckett CS, Dynlacht BD, El-Deiry WS, Elrod JW, Fimia GM, Fulda S, Garcia-Saez AJ, Garg AD, Garrido C, Gavathiotis E, Golstein P, Gottlieb E, Green DR, Greene LA, Gronemeyer H, Gross A, Hajnoczky G, Hardwick JM, Harris IS, Hengartner MO, Hetz C, Ichijo H, Jaattela M, Joseph B, Jost PJ, Juin PP, Kaiser WJ, Karin M, Kaufmann T, Kepp O, Kimchi A, Kitsis RN, Klionsky DJ, Knight RA, Kumar S, Lee SW, Lemasters JJ, Levine B, Linkermann A, Lipton SA, Lockshin RA, Lopez-Otin C, Lowe SW, Luedde T, Lugli E, MacFarlane M, Madeo F, Malewicz M, Malorni W, Manic G, Marine JC, Martin SJ, Martinou JC, Medema JP, Mehlen P, Meier P, Melino S, Miao EA, Molkentin JD, Moll UM, Munoz-Pinedo C, Nagata S, Nunez G, Oberst A, Oren M, Overholtzer M, Pagano M, Panaretakis T, Pasparakis M, Penninger JM, Pereira DM, Pervaiz S, Peter ME, Piacentini M, Pinton P, Prehn JHM, Puthalakath H, Rabinovich GA, Rehm M, Rizzuto R, Rodrigues CMP, Rubinsztein DC, Rudel T, Ryan KM, Sayan E, Scorrano L, Shao F, Shi Y, Silke J, Simon HU, Sistigu A, Stockwell BR, Strasser A, Szabadkai G, Tait SWG, Tang D, Tavernarakis N, Thorburn A, Tsujimoto Y, Turk B, Vanden Berghe T, Vandenabeele P, Vander Heiden MG, Villunger A, Virgin HW, Vousden KH, Vucic D, Wagner EF, Walczak H, Wallach D, Wang Y, Wells JA, Wood W, Yuan J, Zakeri Z, Zhivotovsky B, Zitvogel L, Melino G, Kroemer G (2018) Molecular mechanisms of cell death: recommendations of the nomenclature committee on cell death 2018. Cell Death Differ 25(3):486–541. https://doi.org/10.1038/s41418-017-0012-4
9. Holler N, Zaru R, Micheau O, Thome M, Attinger A, Valitutti S, Bodmer JL, Schneider P, Seed B, Tschopp J (2000) Fas triggers an alternative, caspase-8-independent cell death pathway using the kinase RIP as effector molecule. Nat Immunol 1(6):489–495
10. Degterev A, Hitomi J, Germscheid M, Ch'en IL, Korkina O, Teng X, Abbott D, Cuny GD, Yuan C, Wagner G, Hedrick SM, Gerber SA, Lugovskoy A, Yuan J (2008) Identification of RIP1 kinase as a specific cellular target of necrostatins. Nat Chem Biol 4(5):313–321. https://doi.org/10.1038/nchembio.83
11. Cho YS, Challa S, Moquin D, Genga R, Ray TD, Guildford M, Chan FK (2009) Phosphorylation-driven assembly of the RIP1-RIP3 complex regulates programmed necrosis and virus-induced inflammation. Cell 137(6):1112–1123. https://doi.org/10.1016/j.cell.2009.05.037
12. Zhang DW, Shao J, Lin J, Zhang N, Lu BJ, Lin SC, Dong MQ, Han J (2009) RIP3, an energy metabolism regulator that switches TNF-induced cell death from apoptosis to necrosis. Science 325(5938):332–336. https://doi.org/10.1126/science.1172308
13. He S, Wang L, Miao L, Wang T, Du F, Zhao L, Wang X (2009) Receptor interacting protein kinase-3 determines cellular necrotic response to TNF-alpha. Cell 137(6):1100–1111. https://doi.org/10.1016/j.cell.2009.05.021
14. Sun L, Wang H, Wang Z, He S, Chen S, Liao D, Wang L, Yan J, Liu W, Lei X, Wang X (2012) Mixed lineage kinase domain-like protein mediates necrosis signaling downstream of RIP3 kinase. Cell 148(1–2):213–227. https://doi.org/10.1016/j.cell.2011.11.031
15. Zhao J, Jitkaew S, Cai Z, Choksi S, Li Q, Luo J, Liu ZG (2012) Mixed lineage kinase domain-

like is a key receptor interacting protein 3 downstream component of TNF-induced necrosis. Proc Natl Acad Sci U S A 109(14):5322–5327. https://doi.org/10.1073/pnas.1200012109

16. Cai Z, Jitkaew S, Zhao J, Chiang HC, Choksi S, Liu J, Ward Y, Wu LG, Liu ZG (2014) Plasma membrane translocation of trimerized MLKL protein is required for TNF-induced necroptosis. Nat Cell Biol 16(1):55–65. https://doi.org/10.1038/ncb2883
17. Dondelinger Y, Declercq W, Montessuit S, Roelandt R, Goncalves A, Bruggeman I, Hulpiau P, Weber K, Sehon CA, Marquis RW, Bertin J, Gough PJ, Savvides S, Martinou JC, Bertrand MJ, Vandenabeele P (2014) MLKL compromises plasma membrane integrity by binding to phosphatidylinositol phosphates. Cell Rep 7(4):971–981. https://doi.org/10.1016/j.celrep.2014.04.026
18. Wang H, Sun L, Su L, Rizo J, Liu L, Wang LF, Wang FS, Wang X (2014) Mixed lineage kinase domain-like protein MLKL causes necrotic membrane disruption upon phosphorylation by RIP3. Mol Cell 54(1):133–146. https://doi.org/10.1016/j.molcel.2014.03.003
19. Chen X, Li W, Ren J, Huang D, He WT, Song Y, Yang C, Li W, Zheng X, Chen P, Han J (2014) Translocation of mixed lineage kinase domain-like protein to plasma membrane leads to necrotic cell death. Cell Res 24(1):105–121. https://doi.org/10.1038/cr.2013.171
20. Lin Y, Devin A, Rodriguez Y, Liu ZG (1999) Cleavage of the death domain kinase RIP by caspase-8 prompts TNF-induced apoptosis. Genes Dev 13(19):2514–2526
21. Feng S, Yang Y, Mei Y, Ma L, Zhu DE, Hoti N, Castanares M, Wu M (2007) Cleavage of RIP3 inactivates its caspase-independent apoptosis pathway by removal of kinase domain. Cell Signal 19(10):2056–2067. https://doi.org/10.1016/j.cellsig.2007.05.016
22. O'Donnell MA, Perez-Jimenez E, Oberst A, Ng A, Massoumi R, Xavier R, Green DR, Ting AT (2011) Caspase 8 inhibits programmed necrosis by processing CYLD. Nat Cell Biol 13(12):1437–1442. https://doi.org/10.1038/ncb2362
23. Kaiser WJ, Upton JW, Mocarski ES (2013) Viral modulation of programmed necrosis. Curr Opin Virol 3(3):296–306. https://doi.org/10.1016/j.coviro.2013.05.019
24. Kearney CJ, Martin SJ (2017) An inflammatory perspective on Necroptosis. Mol Cell 65(6):965–973. https://doi.org/10.1016/j.molcel.2017.02.024
25. Gong YN, Guy C, Olauson H, Becker JU, Yang M, Fitzgerald P, Linkermann A, Green DR (2017) ESCRT-III acts downstream of MLKL to regulate Necroptotic cell death and its consequences. Cell 169(2):286–300 e216. https://doi.org/10.1016/j.cell.2017.03.020
26. Orozco S, Yatim N, Werner MR, Tran H, Gunja SY, Tait SW, Albert ML, Green DR, Oberst A (2014) RIPK1 both positively and negatively regulates RIPK3 oligomerization and necroptosis. Cell Death Differ 21(10):1511–1521. https://doi.org/10.1038/cdd.2014.76
27. Newton K, Wickliffe KE, Maltzman A, Dugger DL, Strasser A, Pham VC, Lill JR, Roose-Girma M, Warming S, Solon M, Ngu H, Webster JD, Dixit VM (2016) RIPK1 inhibits ZBP1-driven necroptosis during development. Nature 540(7631):129–133. https://doi.org/10.1038/nature20559
28. Ting AT, Bertrand MJ (2016) More to life than NF-kappaB in TNFR1 signaling. Trends Immunol 37(8):535–545. https://doi.org/10.1016/j.it.2016.06.002
29. Gerlach B, Cordier SM, Schmukle AC, Emmerich CH, Rieser E, Haas TL, Webb AI, Rickard JA, Anderton H, Wong WW, Nachbur U, Gangoda L, Warnken U, Purcell AW, Silke J, Walczak H (2011) Linear ubiquitination prevents inflammation and regulates immune signalling. Nature 471(7340):591–596. https://doi.org/10.1038/nature09816
30. Dondelinger Y, Jouan-Lanhouet S, Divert T, Theatre E, Bertin J, Gough PJ, Giansanti P, Heck AJ, Dejardin E, Vandenabeele P, Bertrand MJ (2015) NF-kappaB-independent role of IKKalpha/IKKbeta in preventing RIPK1 kinase-dependent apoptotic and Necroptotic cell death during TNF signaling. Mol Cell 60(1):63–76. https://doi.org/10.1016/j.molcel.2015.07.032
31. Legarda D, Justus SJ, Ang RL, Rikhi N, Li W, Moran TM, Zhang J, Mizoguchi E, Zelic M, Kelliher MA, Blander JM, Ting AT (2016) CYLD proteolysis protects macrophages from TNF-mediated auto-necroptosis induced by LPS and licensed by type I IFN. Cell Rep 15(11):2449–2461. https://doi.org/10.1016/j.celrep.2016.05.032
32. Vandenabeele P, Grootjans S, Callewaert N, Takahashi N (2013) Necrostatin-1 blocks both RIPK1 and IDO: consequences for the study of cell death in experimental disease models. Cell Death Differ 20(2):185–187. https://doi.org/10.1038/cdd.2012.151
33. Wang L, Du F, Wang X (2008) TNF-alpha induces two distinct caspase-8 activation pathways. Cell 133(4):693–703
34. Najjar M, Saleh D, Zelic M, Nogusa S, Shah S, Tai A, Finger JN, Polykratis A, Gough PJ, Bertin J, Whalen M, Pasparakis M, Balachandran S, Kelliher M, Poltorak A, Degterev A (2016)

RIPK1 and RIPK3 kinases promote cell-death-independent inflammation by toll-like receptor 4. Immunity 45(1):46–59. https://doi.org/10.1016/j.immuni.2016.06.007

35. Daniels BP, Snyder AG, Olsen TM, Orozco S, Oguin TH 3rd, Tait SWG, Martinez J, Gale M Jr, Loo YM, Oberst A (2017) RIPK3 restricts viral pathogenesis via cell death-independent Neuroinflammation. Cell 169(2):301–313 e311. https://doi.org/10.1016/j.cell.2017.03.011
36. Moriwaki K, Balaji S, McQuade T, Malhotra N, Kang J, Chan FK (2014) The necroptosis adaptor RIPK3 promotes injury-induced cytokine expression and tissue repair. Immunity 41(4):567–578. https://doi.org/10.1016/j.immuni.2014.09.016
37. Newton K, Dugger DL, Maltzman A, Greve JM, Hedehus M, Martin-McNulty B, Carano RA, Cao TC, van Bruggen N, Bernstein L, Lee WP, Wu X, DeVoss J, Zhang J, Jeet S, Peng I, McKenzie BS, Roose-Girma M, Caplazi P, Diehl L, Webster JD, Vucic D (2016) RIPK3 deficiency or catalytically inactive RIPK1 provides greater benefit than MLKL deficiency in mouse models of inflammation and tissue injury. Cell Death Differ 23(9):1565–1576. https://doi.org/10.1038/cdd.2016.46

Chapter 2

Loss-of-Function RNAi Screen to Identify Necrosis-Signaling Molecules

David Mark Moquin and Francis Ka-Ming Chan

Abstract

Over the recent years, genome-wide RNA interference (RNAi) library screens have been instrumental in the identification of key regulators of various biological pathways. The prolific use of this technique is attributed to its amenability to a high-throughput format. Here, we present the step-by-step method to conduct a siRNA screen to identify genes involved in necroptosis, a nonapoptotic form of proinflammatory cell death. The method described here uses MTS cell proliferation assay to measure necroptosis, which is compatible with high-throughput format screening on multiwell microtiter plates. This ensures that the screen can be performed in a timely and efficient manner.

Key words Necroptosis, Programmed necrosis, Cell death, Caspase-independent, TNFα, Transfection, siRNA screen

1 Introduction

Necroptosis or programmed necrosis is a caspase-independent form of cell death that promotes inflammation [1]. Some of the key regulators involved in this process, such as the serine/threonine kinase receptor interacting protein kinase 3 (RIPK3) [2, 3], the tumor suppressor/deubiquitinase CYLD [4, 5], and the downstream effector mixed lineage kinase domain-like (MLKL) [6], were initially identified through screening siRNA or shRNA libraries. This highlights the utility of high throughput-based screening assays to identify regulators of necroptosis.

Synthetic small interfering RNAs (siRNAs) are double stranded, 21–25 nucleotides long. The argonaute proteins within the RNA-induced silencing complex (RISC) recognize the 3′ overhangs of siRNAs. The RISC scans mRNAs for complementarity to the antisense strand of the siRNA to inhibit translation or cause degradation of the target mRNAs [7]. siRNAs are ideal for transient knockdown of genes, as their small size allows for robust

Adrian T. Ting (ed.), *Programmed Necrosis: Methods and Protocols*, Methods in Molecular Biology, vol. 1857,
https://doi.org/10.1007/978-1-4939-8754-2_2, © Springer Science+Business Media, LLC, part of Springer Nature 2018

transfection efficiency. They also bypass some of the processing steps required for short hairpin RNAs (shRNAs).

Introduction of siRNAs into cells is achieved by transfection with a cationic lipid transfection reagent. Using this protocol, efficient knockdown can be achieved between 48–72 h post transfection. The cells used in the primary screen are Fas-Associated protein with Death Domain (FADD)-deficient Jurkat cells that express Tumor Necrosis Factor Receptor 2 (TNFR2$^+$), a cell line that is highly sensitive to TNF-induced necroptosis [8]. These cells are an excellent choice for identifying regulators of necroptosis as they readily undergo TNF-induced necroptosis without caspase inhibitors. This will significantly reduce the cost and time for the screen.

Necroptosis will be measured using the CellTiter 96® Aqueous One Solution Cell Proliferation Assay from Promega, which measures cellular dehydrogenase activity and is a direct correlate of cell viability. Cellular dehydrogenase activity converts tetrazolium compound into a formazen product, which can be measured in a microtiter plate reader in a high-throughput manner.

2 Materials

2.1 Disposable Tissue Culture Supplies

1. V-bottom 96-well plates.
2. RNase-free filter pipette tips, 10 μL, 20 μL, and 200 μL.
3. Serological pipettes, 5 and 10 mL.
4. MicroAmp adhesive film for microtiter plate.
5. Reagent reservoir for multichannel pipettor.

2.2 siRNA Transfection

1. Human siRNA library (*see* **Note 1**).
2. 1× siRNA annealing buffer: 30 mM HEPES [pH 7.4], 100 mM potassium acetate, 2 mM magnesium acetate.
3. Qiagen HiPerfect transfection reagent or equivalent.
4. Serum-free RPMI 1640 medium with no supplement.
5. Negative control nonsilencing siRNA (NS siRNA).
6. Positive control siRNA against human RIPK1 (5’-rUrGrCrArGrUrCrUrCrUrUrCrArArCrUrUrGrAdTdT-3′, 5′-rUrGrCrUrCrUrUrCrArUrUrArUrUrCrArGrUrUrUrGrCrUrCrCrArC-3′).

2.3 Necroptosis Assay

1. TNFR2$^+$FADD-deficient Jurkat cells [8].
2. Recombinant human TNFα (rhTNFα).
3. zVAD-fmk.
4. Staurosporine.
5. CellTiter 96® Aqueous One Solution Cell Proliferation Assay (Promega).

6. Complete RPMI 1640 medium: RPMI 1640 medium, 10% fetal calf serum, 2 mM glutamine, 100 units/mL penicillin, 100 μg/mL streptomycin.

3 Methods

3.1 Preparation of siRNA Transfection Mixture (See Note 2)

1. Warm serum-free RPMI 1640 medium in 37 °C water bath. Retrieve pooled siRNA library from −80 °C freezer and thaw it at room temperature for 5 min. Centrifuge the plate containing the siRNA stocks at 700 × *g* for 5 min using a swinging bucket centrifuge.
2. Carefully remove the plastic seal from the plate (*see* **Note 3**). If the siRNA is lyophilized, resuspend the siRNA in 1× annealing buffer to make a stock concentration of 10 μM.
3. Aliquot 1.5 μL of the pooled siRNA into a new 96-well plate (*see* Fig. 1 for a typical layout of the plate). This will yield a 150 nM final concentration for the siRNA (*see* **Note 4**). Reseal

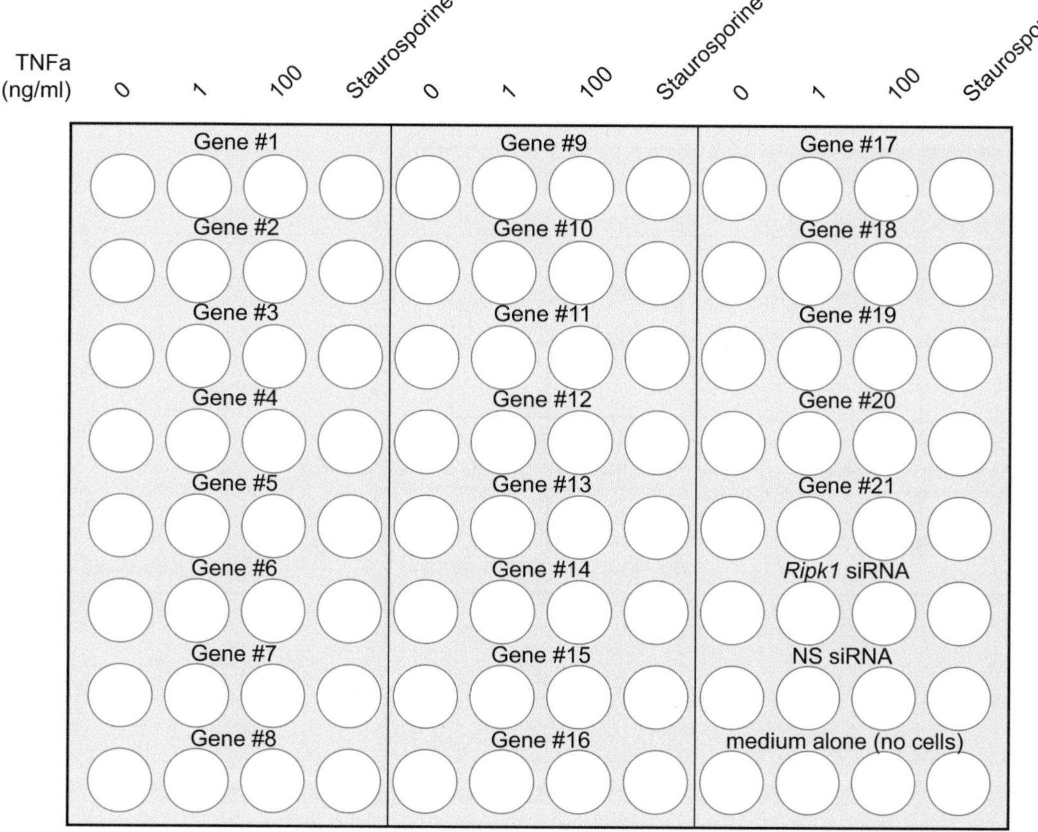

Fig. 1 Layout of representative 96-well plate for siRNA screening. In this format, up to 21 genes can be screened on a single 96-well microtiter plate. Note that the last four wells contain only the culture medium without any cells. They are used to subtract the background absorbance in the CellTiter 96® Aqueous One Solution Cell Proliferation assay

the plate containing the stock siRNAs with a new adhesive plastic film and return the plate to storage at −80 °C.

4. Prepare a master mix of HiPerfect transfection reagent diluted in serum-free RPMI 1640. Use 0.75 μL HiPerfect transfection reagent and 24.25 μL serum-free RPMI 1640 medium per well. For each 96-well plate, add 75 μL of HiPerfect reagent dropwise to 2.425 mL of serum-free RPMI 1640 medium in a multichannel pipette reagent reservoir. Mix gently by pipetting up and down with a 5-mL serological pipette.
5. Transfer 25 μL of HiPerfect transfection reagent diluted in serum-free RPMI 1640 to the 96-well plate containing siRNA. This can be achieved using a multichannel pipetman (*see* Fig. 2 for workflow).
6. Incubate at room temperature for 5–10 min.
7. Resuspend FADD-deficient $TNFR2^+$ Jurkat cells at 8.57×10^4 cells per mL Complete RPMI 1640 medium. Use a multichan-

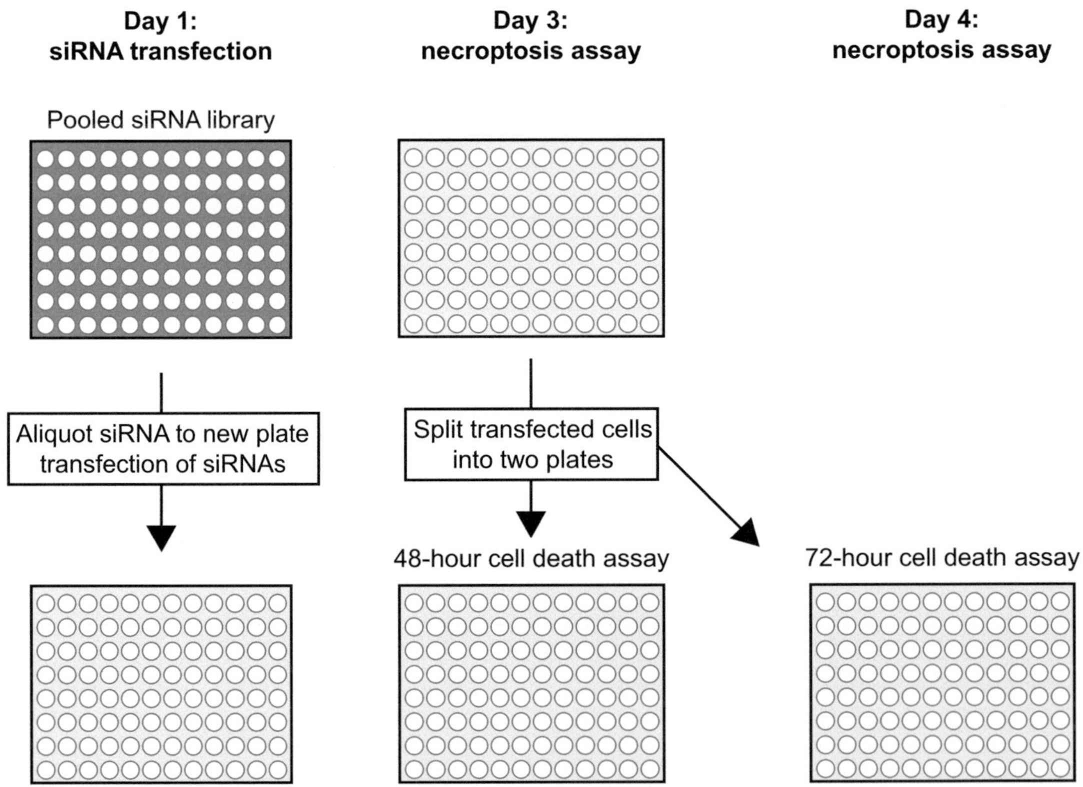

Fig. 2 Schematic diagram of the workflow of the siRNA library screen. 48 h after siRNA transfection, half the cells in each well are transferred to a corresponding new 96-well plate. Cells in the original plate are then treated with the death ligands. RNA interference proceeds for another 24 h in the duplicate plate before an identical cell death assay is initiated

nel pipetman to dispense 175 μL of cell suspension (1.5×10^4 cells) to each well containing the siRNA/HiPerfect mixture. Mix by pipetting up and down 4 times (*see* **Notes 5** and **6**). Return the plate to 37 °C incubator with 5% CO_2.

3.2 Induction of Cell Death

1. After 48 h, transfer 100 μL of cells to another 96-well plate.
2. Use one of the plates for the 48-h post-transfection necroptosis assay. Return the second plate to the 37 °C incubator with 5% CO_2 for the 72 h post-transfection necroptosis assay.
3. For each 96-well plate to be assayed, prepare four tubes containing 3 mL of complete RPMI 1640 medium. To the first two tubes, add rhTNFα (*see* **Note 7**) to final concentrations of 200 ng/mL and 2 ng/mL respectively. To the third tube, add 1 μg/mL staurosporine (*see* **Note 8**). This is the apoptosis control. The fourth tube contains only complete RPMI 1640 medium without cell death ligands. This is the untreated control.
4. Transfer 100 μL of complete RPMI 1640 or the diluted cell death ligands to the indicated samples. *See* Fig. 1 for the layout of the stimulations. Incubate for 6–14 h at 37 °C.

3.3 Measurement of Cell Death

1. Thaw CellTiter 96® Aqueous One Solution Cell Proliferation assay (*see* **Note 9**) in a 37 °C water bath for 10 min.
2. Transfer 40 μL of CellTiter 96® Aqueous One Solution to each well. Dispense 4 mL of CellTiter 96® Aqueous One Solution to a multichannel reagent reservoir. Using a multichannel pipetman, add 40 μL of CellTiter 96® Aqueous One Solution to each well.
3. Incubate for 1–4 h at 5% CO_2 37 °C incubator.
4. Measure the absorbance at 490 nm (Ab_{490}) using a plate reader (*see* **Notes 10–12**).
5. Calculate the absorbance values of the wells by subtracting the absorbance values of wells that has no cells (Fig. 1).
6. Calculate the percentage cell loss for each siRNA transfected sample using the formula:

   % Cell Loss = (1 − (Ab_{490} treated sample/Ab_{490} untreated sample) × 100%
7. Calculate the average percentage cell loss from samples transfected with different siRNAs. Select clones that show protection against TNF-induced necroptosis for further validation (*see* **Note 13**).

3.4 Validation of Putative Positives

1. Validate putative positive clones for protection against TNF-induced necroptosis by transfecting cells with individual siRNAs. Those clones that show protection with two or more siRNAs are subjected to further analysis (*see* **Note 14**).

2. Genes that show protection against TNF-induced necroptosis by at least two different siRNAs will be further tested for protection in caspase 8-deficient Jurkat cells and wild-type Jurkat cells expressing TNFR2. To induce necroptosis in wild-type TNFR2$^+$ Jurkat cells [9], use TNF and the pancaspase inhibitor zVAD-fmk (10 μM) (*see* **Note 15**).
3. Validate the knockdown efficiency for each hit via quantitative RT-PCR and by Western blot (*see* **Note 16**).

4 Notes

1. There are many commercially available siRNA libraries. You can customize your library according to your need (e.g., screening only subset of genes). We typically use libraries in which each gene is targeted by four distinct siRNA oligonucleotides. However, we have also used libraries with only two distinct siRNAs against each gene. Typically, we prefer to use libraries in which the individual siRNAs are assembled on 96-well plate separately. This is particularly useful for subsequent validation of putative targets using individual rather than pooled siRNAs.
2. Always ensure that the work area is RNase free. This can be accomplished by wiping down the work area, pipetmen, and other equipment with RNaseZap® or equivalent. It is also advisable to use filter barrier tips and designate pipette tip boxes specifically for the siRNA screen to minimize cross-contamination from other experiments. Always wear gloves when working with siRNA to minimize contamination with RNases.
3. We typically ordered small quantities of siRNA for the screen (e.g., 0.25 nmoles). Care should be taken to not spill any of the siRNA while removing the plastic seal from the plate. Only open the sealed siRNA plate in a biosafety hood.
4. The siRNA concentration used is critical for success of the screen. Optimal siRNA concentration can vary due to siRNA modifications, the transfection reagent, and cell line used. High concentration of siRNA can increase the possibility of off-target effects. Optimization of transfection conditions is recommended. This can be achieved using siRNA against a known cellular gene.
5. If handling multiple plates in the same experiment, make sure that cells do not settle to the bottom of the reservoir over time. Resuspend the cells periodically to ensure an equal number of cells are plated for each well.

6. When using a multichannel pipette always visually examine the pipette tips to ensure that equal volume is being dispensed into each well.
7. The activity of rhTNFα can vary depending on the vendor and the batch. Therefore, it is prudent to conduct a pilot cell death assay to determine the optimal dose, incubation time and the cell type used. Using TNF doses near the ED_{50} is recommended as it allows potential identification of genes that activates as well as inhibits TNF-induced necroptosis.
8. Staurosporine is a kinase inhibitor that induces apoptosis. Inclusion of staurosporine in the screen allows identification of specific regulators of necroptosis from those that controls cell viability in general.
9. The MTS assay, which is what the CellTiter 96® Aqueous One Solution Cell Proliferation assay is, measures mitochondrial metabolic activity. Although it does not directly measure cell death, it is generally a reliable method for measuring the number of viable cells in a culture. Other microtiter plate-based assays, such as release of lactate dehydrogenase (LDH) [10], or high-throughput FACS-based assays, can also be used.
10. Bubbles in the wells can interfere with absorbance readings. Make sure to purge the air bubbles before measurement. This can be achieved using a 22-gauge needle.
11. The rate of conversion of the MTS absorbance varies depending on the cell type and the number of cells used. As such, it is recommended that multiple measurements be taken throughout the incubation period to ensure that readings fall within the linear range.
12. The formation of the formazen product for the viability assay can be terminated by adding 50 μL of 10% SDS per well. This allows measurement up to 18 h later if the plate is stored at room temperature with humidity. Make sure to protect the plate from light exposure.
13. We typically calculate the average percentage cell loss from at least 20 different samples. We then use a cutoff of three standard deviations for selecting putative positive clones. However, a lower cutoff can be used to increase the number of putative positives to screen.
14. Given the well-documented possibility of siRNA off-target effects, it is important to validate protection against necroptosis by multiple siRNAs. Validation with multiple siRNAs increases the confidence that the protection is not caused by off-target effects.
15. Validating protection against necroptosis in these two additional cell lines provides additional assurance that the protec-

tive effect seen with any putative positive clones is authentic. Other necroptosis sensitive cell lines, such as L929 [4] or HT-29 [11], can also be used.

16. Knockdown efficiency by the individual siRNAs should be confirmed by Q-PCR. Where the antibody reagent is available, Western blot should also be performed to validate knockdown efficiency.

References

1. Chan FK, Luz NF, Moriwaki K (2015) Programmed necrosis in the cross talk of cell death and inflammation. Annu Rev Immunol 33:79–106
2. Cho YS, Challa S, Moquin D, Genga R, Ray TD, Guildford M, Chan FK (2009) Phosphorylation-driven assembly of the RIP1-RIP3 complex regulates programmed necrosis and virus-induced inflammation. Cell 137:1112–1123
3. He S, Wang L, Miao L, Wang T, Du F, Zhao L, Wang X (2009) Receptor interacting protein kinase-3 determines cellular necrotic response to TNF-alpha. Cell 137:1100–1111
4. Hitomi J, Christofferson DE, Ng A, Yao J, Degterev A, Xavier RJ, Yuan J (2008) Identification of a molecular signaling network that regulates a cellular necrotic cell death pathway. Cell 135:1311–1323
5. Moquin DM, McQuade T, Chan FK (2013) CYLD deubiquitinates RIP1 in the TNFalpha-induced necrosome to facilitate kinase activation and programmed necrosis. PLoS One 8:e76841
6. Zhao J, Jitkaew S, Cai Z, Choksi S, Li Q, Luo J, Liu ZG (2012) Mixed lineage kinase domain-like is a key receptor interacting protein 3 downstream component of TNF-induced necrosis. Proc Natl Acad Sci U S A 109:5322–5327
7. Wilson RC, Doudna JA (2013) Molecular mechanisms of RNA interference. Annu Rev Biophys 42:217–239
8. Chan FK, Shisler J, Bixby JG, Felices M, Zheng L, Appel M, Orenstein J, Moss B, Lenardo MJ (2003) A role for tumor necrosis factor receptor-2 and receptor-interacting protein in programmed necrosis and antiviral responses. J Biol Chem 278:51613–51621
9. Chan FK, Lenardo MJ (2000) A crucial role for p80 TNF-R2 in amplifying p60 TNF-R1 apoptosis signals in T lymphocytes. Eur J Immunol 30:652–660
10. Chan FK, Moriwaki K, De Rosa MJ (2013) Detection of necrosis by release of lactate dehydrogenase activity. Methods Mol Biol 979:65–70
11. Wang L, Du F, Wang X (2008) TNF-alpha induces two distinct caspase-8 activation pathways. Cell 133:693–703

Chapter 3

Chemical Library Screens to Identify Pharmacological Modulators of Necroptosis

Danish Saleh and Alexei Degterev

Abstract

Necroptosis is mediated by the formation of the detergent-insoluble necrosome complex between Ser/Thr kinases RIPK1 and RIPK3, which mediates RIPK3-dependent phosphorylation and activation of the critical necroptosis effector MLKL. Small molecule screens have been instrumental in the development of new chemical probes for this pathway. In this chapter, we describe several cellular assays that are readily amendable for the identification of new modulators of necroptosis as well as secondary assays to facilitate initial characterization of the mode of activity of small molecule hits.

Key words Necroptosis, Apoptosis, Necrosome, Complex 2b, RIPK1, RIPK3, MLKL

1 Introduction

Necroptosis is a process of regulated necrotic cell death, which has attracted major interest due to its role in a broad range of pathologic settings [1–3]. Availability of small molecule inhibitors of a number of key players in this pathway, including RIPK1, RIPK3, MLKL, caspase-8, and inhibitor of apoptosis (IAP) proteins, has been instrumental in both defining the central mechanism of necroptosis as well as its contributions to disease pathophysiology [4]. These molecules were identified through a variety of approaches including cell based screens [5–7], in vitro screens utilizing purified recombinant factors [8–11], and identification of new anti-necroptotic activities of the existing inhibitors, e.g., panspecific kinase inhibitors or Hsp90 inhibitors [12–15]. Cell based screens may offer significant benefits, including unbiased selection of the molecules with the preferred mode of activity, i.e., leading to the maximal functional benefit, exclusion of the molecules with on-target or off-target toxicities and suboptimal cell permeability, and the opportunity to elucidate new regulatory modalities in a pathway. It is also becoming clear that multiple players in necroptosis possess functions that extend beyond necroptosis, including

Adrian T. Ting (ed.), *Programmed Necrosis: Methods and Protocols*, Methods in Molecular Biology, vol. 1857,
https://doi.org/10.1007/978-1-4939-8754-2_3,

regulation of apoptosis and inflammation, which may also contribute to pathologies in a context-specific manner. These responses can be mimicked in cells, allowing for selection of lead molecules and/or chemical probes, specifically affecting new responses, mediated by necroptotic factors. However, cell-based screens offer just the first step that should be followed by oftentimes elaborate target identification and validation studies. In this chapter, we describe several cellular models that are well suited for primary screening as well as several simple and robust secondary assays that may facilitate analysis of the points of action, if not the targets, of the newly identified inhibitors.

2 Materials

2.1 Inducers of RIPK Signaling

1. Human and mouse TNFα (R&D Systems and Peprotech). Lyophylized proteins are dissolved in water at 100 ng/mL and stored in −80 °C.
2. Other TNF family members, such as FasL and TRAIL. Lyophylized proteins are dissolved in water at 100 ng/mL and stored in −80 °C.
3. Lipopolysaccharide (LPS) from multiple gram-negative bacterial species (Sigma or Invivogen). LPS is dissolved in water at 1 mg/mL and stored in −20 °C. Other TLR agonists can also be used [16, 17], although details of the regulation may differ.
4. Interferon(IFN)-beta and interferon(IFN)-gamma (PBL Assay Science).
5. Pancaspase inhibitors: zVAD.fmk, IDN-6556/Emricasan or qVD-OPh (ApexBio). Prepare as 10–100 mM stocks in DMSO and store in −20 °C.
6. TAK1 inhibitor: 5z-7-oxozeaenol (Sigma-Aldrich, ApexBio, Tocris Bioscience). Prepare as 1 mM stocks in DMSO and store in −20 °C.
7. SMAC mimetics: SM164 (Sigma-Aldrich, ApexBio, Tocris Bioscience). Prepare as 1 mM stocks in DMSO, and store in −20 °C.
8. Cycloheximide (CHX) (Sigma-Aldrich). Dissolve in water at 10 mg/mL and store at −20 °C.

2.2 Cells and Tissue Cultures

Multiple lines of human and mouse cells have been used widely to study mechanisms of necroptosis (*see* **Note 1**), and are well suited for screening due to efficient activation of this form of cell death, rapid doubling times, and growth under standard media conditions (DMEM or RPMI1640 supplemented with 10% fetal bovine serum and 1% antibiotic–antimycotic mix).

1. Human Jurkat T cells.
2. Human U937 monocytic cells.
3. Human HT29 adenocarcinoma cells.
4. Mouse L929 fibrosarcoma cells.
5. Mouse RAW264.7 macrophages.
6. Spontaneously immortalized or SV40-transformed mouse embryonic fibroblasts (MEFs).
7. Complete RPMI1640 medium: RPMI1640, 10% FBS (can be substituted with FetalPlex, GeminiBio #100-602) and 1% antibiotic–antimycotic mix.
8. Complete DMEM medium: DMEM, 10% FBS, and 1% antibiotic–antimycotic mix.

2.3 Antibodies for Necrosome Analyses

1. Anti-RIPK1 rabbit monoclonal D94C12 (Cell Signaling #3493).
2. Anti-phospho-RIPK1 (S166) rabbit polyclonal (Cell Signaling #31122).
3. Anti-RIPK3 rabbit polyclonal (ProSci #2283).
4. Anti-MLKL rabbit polyclonal (Biorbyt #orb32399).
5. Anti-phospho-MLKL (S345) rabbit monoclonal EPR9515(2) (Abcam #ab196436).
6. Rabbit IgG (Santa Cruz). Use as negative control for immunoprecipitation.
7. Pierce protein A-conjugated magnetic beads (ThermoFisher Scientific).

2.4 Other Materials

1. Recombinant RIPK1 and RIPK3 proteins [14].
2. Kinase reaction buffer: 50 mM HEPES, pH 7.5, 50 mM NaCl, 30 mM $MgCl_2$, 1 mM DTT, 0.05% bovine serum albumin (BSA), and 0.02% CHAPS.
3. Necrosome Lysis Buffer: 0.05% Triton X-100, 150 mM NaCl, 20 mM Tris–HCl (pH 7.5), 10% glycerol (optional), 1 mM EDTA, 3 mM NaF, 1 mM beta-glycerophosphate, 1 mM sodium orthovanadate, 5 μM idoacetamide, 2 μM N-ethylmaleimide, 5 μg/mL PMSF, 1 μg/mL leupeptin, 1 μg/mL pepstatin, and 1 μg/mL aprotinin.
4. CellTiter-Glo® Luminescent Cell Viability Assay (Promega).
5. RS repeat peptide (SignalChem).
6. ADP-Glo Kinase Assay (Promega).

3 Methods

3.1 Activation of Necroptosis by TNFα in Jurkat Cells

FADD-deficient Jurkat cells provide a robust and convenient system to identify modulators of necroptosis (*see* **Note 2**). Treatment of the cells with 10 ng/mL human TNFα for 18–24 h typically leads to 60–80% loss of viability measured by CellTiter-Glo assay.

1. Cells are maintained in complete RPMI1640 medium at densities of 1×10^5–1×10^6 cells/mL. Cells can typically be maintained in culture for 3–4 weeks without a major decrease in the sensitivity to necroptosis.
2. Cells are plated in the fresh complete RPMI1640 medium at the density of 5×10^4 cells/mL in 100 μL in 96-well plates (white or white with clear bottoms). For 384-well plates, cells are seeded in 40 μL of medium (2×10^4 cells/well).
3. Small molecules are added from DMSO stocks using available liquid handling equipment, but final DMSO concentration in the media should not exceed 0.5–1%.
4. Human TNFα is diluted in PBS and added to the wells to achieve a final concentration of 10 ng/mL (*see* **Note 3**).
5. Each test plate should include triplicate wells of cells treated with just DMSO (100% viability control) and TNFα/DMSO (maximal cell death control). It is also recommended to include a positive control compound, such as RIPK1 inhibitor Nec-1s or RIPK3 inhibitor GSK'872 at 5 μM, which may also serve as interplate standards.
6. Cells are maintained in the humidified 5% CO_2 incubator for 18–24 h.
7. Cells are removed from the incubator, and 15 μL of CellTiter-Glo (Promega) Cell Viability reagent is added per 100 μL of medium.
8. Plates are incubated (preferably on a shaker) at room temperature for 10 min and luminescence signals are determined using appropriate luminescence plate-reader (typical integration time 0.3–1 s).
9. Viability in each well is determined as the ratio of the values in test wells to DMSO-only positive control (maximal cell death) wells. The viability of FADD-deficient Jurkat T cells treated with TNF is shown in Fig. 1a.

3.2 Activation of RIPK3-Dependent Necroptosis by Interferons in *Ripk1*$^{-/-}$ MEFs

In a number of instances, upstream necroptotic signals have been shown to bypass RIPK1 and directly engage RIPK3. This was reported, for instance, upon the activation of the viral sensor DAI, treatment of human epithelial SVEC4-10 or mouse fibroblast NIH3T3 cells with TLR3/4 agonists/zVAD, and in response to IFNβ or IFNγ in *Ripk1*$^{-/-}$ MEFs [14, 16, 18, 19]. The use of these

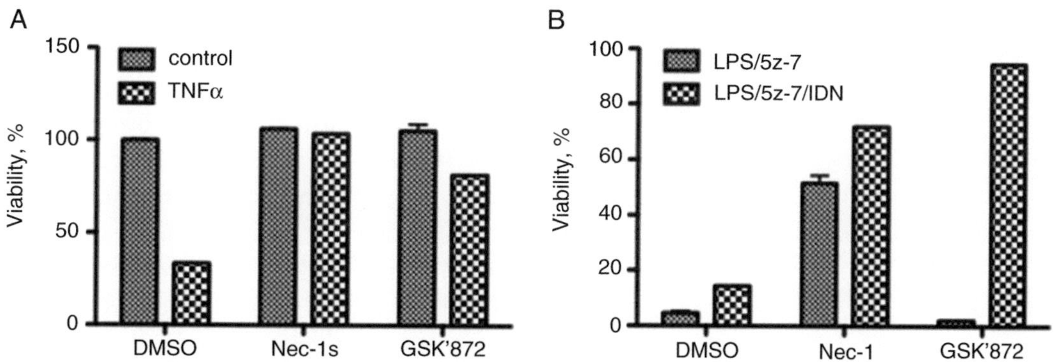

Fig. 1 Cellular assays for RIPK-dependent cell death. (**a**) FADD-deficient Jurkat cells were treated with 10 ng/mL TNFα. (**b**) RAW264.7 cells were treated with 100 ng/mL *E. coli* LPS, 300 nM 5z-7-oxozeaenol, and 10 μM IDN-6556. In both cases, cells were also treated with 5 μM Nec-1s and GSK'872. CellTiter-Glo Viability assay was performed after 24 h treatment

model systems or dimerizable RIPK3/MLKL, described by Rodriguez and Green in Chapter 7 of this same volume, will bias selection toward inhibition of RIPK3/MLKL steps in the pathway.

1. *Ripk1*$^{-/-}$ MEFs are maintained in complete DMEM medium.
2. Cells are seeded overnight into 96-well plates (white or white with clear bottoms) at 1×10^4 cells in 100 μL media (2×10^3 cells/40 μL media in 384-well plates).
3. Cells are treated with inhibitors, followed by the addition of 10 ng/mL IFNγ[3].
4. Each test plate should include triplicate wells of cells treated with just DMSO (100% viability control) and IFNγ/DMSO (maximal cell death control). It is also recommended to include wells treated with IFNγ/GSK'872 at 5 μM to serve as a positive control and an interplate standard.
5. Cells are maintained in the 5% CO_2 humidified incubator for 48 h.
6. CellTiter-Glo assay is performed as in Subheading 3.1.

3.3 Alternative Activation of RIPK1-Dependent Apoptosis vs. RIPK1/RIPK3-Dependent Necroptosis in RAW264.7 Cells

RAW264.7 macrophages and MEF fibroblasts undergo apoptosis that requires the catalytic activity of RIPK1 in response to LPS or TNFα, respectively, in the presence of 5z-7-oxozeaenol (or SM164) (Fig. 1b). Inhibition of caspase-8 by zVAD.fmk (or other caspase inhibitors) leads to alternative activation of necroptosis, which is efficiently blocked by RIPK3 inhibitors (and less efficiently by RIPK1 inhibitors) [14].

1. RAW264.7 cells are maintained in DMEM supplemented with 10% FBS and 1% antibiotic–antimycotic mix.

2. Cells are seeded overnight into 96-well plates (white or white with clear bottoms) at 1×10^4 cells in 100 μL media (2×10^3 cells/40 μL media in 384-well plates) in the medium supplemented with 100 nM 5z-7-oxozeaenol (or SM164) and 20–50 μM zVAD (optional).
3. Cells are treated with inhibitors, followed by the addition of 10 ng/mL LPS.
4. Each test plate should include triplicate wells of cells treated with just DMSO (100% viability control) and LPS/5z-7/(zVAD)/DMSO (maximal cell death control). It is also recommended to include wells treated with LPS/5z-7/(zVAD)/Nec-1s(GSK'872) to serve as a positive control and an interplate standard.
5. Cells are maintained in the 5% CO_2 humidified incubator for 8–16 h.
6. CellTiter-Glo assay is performed as in Subheading 3.1. Figure 1b shows the viability of RAW264.7 cells treated with necroptosis inducers in the presence or absence of necroptosis inhibitors.

3.4 In Vitro Kinase Assays for RIPK1 and RIPK3

Because catalytic activities of RIPK1 and RIPK3 play central roles in necroptosis, many inhibitors identified in cell based screen are likely to directly target these kinases. Thus, secondary in vitro assays against these two kinases are useful for the initial evaluation of the screening hits.

1. Recombinant RIPK1 and RIPK3 kinases are expressed in Sf9 cells using corresponding baculoviruses as described in [14].
2. Proteins (20 ng) are diluted in the kinase reaction buffer.
3. Diluted proteins are added to the low volume white 384-well plates (2 μL/well).
4. Inhibitors in DMSO are diluted in the kinase reaction buffer (to obtain a final DMSO concentration of 25%). 1 μL is added to each well and plates are incubated for 5 min at room temperature.
5. Reactions are initiated by the addition of 2 μL of 100 μM ATP and 1 mg/mL RS repeat peptide in the kinase reaction buffer.
6. Each test plate should include triplicate wells of RIPK/DMSO (100% activity control) and reaction buffer/DMSO (background signal). It is also recommended to include a positive control compound, such as RIPK1 inhibitor Nec-1s or RIPK3 inhibitor GSK'872 at 5 μM, to serve as a positive control and an interplate standard.
7. Plates are covered with plastic/metal seals, and incubated at room temperature for 3–4 h.

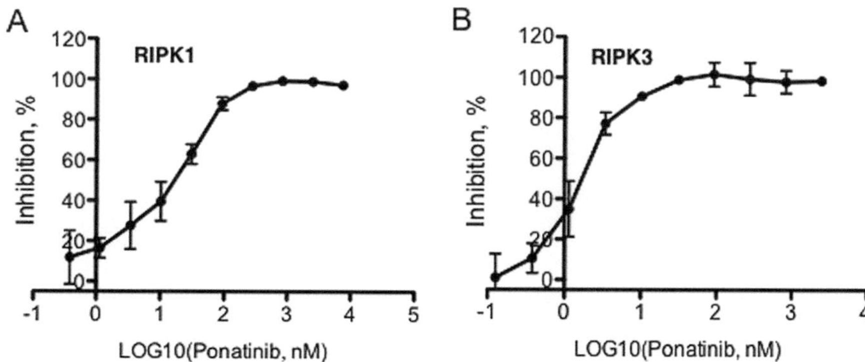

Fig. 2 Examples of in vitro recombinant RIPK1 and RIPK3 kinase inhibition curves. ADP-Glo assays were performed using a dose range of ponatinib, which inhibits both kinases [14]

8. Reactions are stopped by the addition of 5 μL of ADP-Glo reagent and ADP conversion reaction is performed for 40 min at room temperature.
9. Luminescence signal is generated by the addition of 10 μL of Kinase detection reagent for 30 min at room temperature.
10. Luminescence signals are determined using appropriate luminescence platereader (typical integration time 0.3–1 s).
11. To calculate percent inhibition, average background signal is subtracted from test well and maximal signal wells. Inhibition, % = (1− (test signal/maximal signal)) × 100. IC_{50} values are calculated based on a dose range of inhibitor concentrations using nonlinear regression in GraphPad Prism software. An example showing the inhibition of RIPK1 and RIPK3 by ponatinib is shown in Fig. 2.

3.5 Necrosome Isolation by Differential Centrifugation

Activation of RIPK1 and RIPK3 and induction of necroptosis is associated with the central step of the formation of amyloid-like complexes of RIPK1 and RIPK3, also known as necrosomes, which are insoluble in mild detergents [20–23]. Treatment of RAW264.7 cells with LPS (10 ng/mL) and a pancaspase inhibitor (50 μM zVAD.fmk or 10 μM IDN-6556) promotes necrosome formation, evaluated as coenrichment of RIPK1, RIPK3, and MLKL in the detergent insoluble cellular fractions.

1. RAW264.7 cells are seeded overnight in 10 cm^2 tissue-culture treated plates at 1 × 10^7 cells/plate in 10 mL of complete DMEM medium (*see* **Note 4**).
2. Cells are treated with LPS (10 ng/mL) and zVAD.fmk (50 μM) or IDN-6556 (10 μM) +/− Nec-1s (30 μM) as a control for the specific activation of RIPK1 kinase-dependent necroptosis for 3–4 h (*see* **Note 5**). Plates are maintained in an incubator at 37 °C with 5% CO_2 during treatment.

3. Following the treatment period, adherent and nonadherent cells are harvested using the Necrosome Lysis Buffer. Cells are harvested by removing culture media and adding 300–500 μL of lysis buffer to each plate. Adherent cells are lysed and collected by mechanical scraping directly from the plate while maintained on ice. Nonadherent cells are collected from the cell culture media (*see* **Note 6**).
4. Lysates are flash-frozen on dry ice, and then thawed at room temperature or on ice.
5. Once thawed, samples are briefly vortexed (5–10 s, setting 4.5), and centrifuged at 1000 × *g* at 4 °C in a tabletop centrifuge for 15 min to pellet nuclei.
6. Supernatants (S1) are collected, and protein concentrations are normalized across samples using Pierce 660 nM protein assay or similar.
7. S1 samples are centrifuged at 34,400 × *g* at 4 °C for 15 min to precipitate Triton-insoluble pellet.
8. The resultant supernatant (S2; Triton-soluble fraction) is collected and diluted in 4× Laemmli sample buffer for SDS-PAGE/Western analysis (*see* **Note 7**).
9. Once S2 sample has been collected, residual supernatant is discarded (*see* **Note 8**). The resultant pellet is washed by suspending in 300 μL of Necrosome Lysis Buffer and centrifugation at 34,400 × *g* at 4 °C for 15 min (*see* **Note 9**).
10. Following centrifugation, the supernatant is discarded (*see* **Note 10**).
11. Detergent insoluble pellets are boiled in 50–60 μL of 1× Laemmli sample buffer for 5–10 min to prepare SDS-PAGE samples (*see* **Note 11**).
12. Coenrichment of detergent-insoluble RIPK1, pRIPK1, RIPK3, MLKL, or pMLKL in LPS/zVAD (or LPS/IDN-6556) treated cells compared to LPS/zVAD/Nec-1s or LPS/IDN-6556/Nec-1s is analyzed by Western blotting. A flowchart of the centrifugation procedure and a blot showing necrosome components isolated from RAW264.7 cells treated with LPS + IDN-6556 are shown in Fig. 3.

3.6 Necrosome Isolation Using a Sucrose Gradient and Velocity Sedimentation

In lieu of differential centrifugation, velocity sedimentation along a 10–50% linear sucrose gradient can be employed to further examine necrosome formation (*see* **Note 12**).

1. RAW264.7 cells are seeded in 3–4 10 cm^2 dishes at 12–15 × 10^6 cells/plate per experimental condition in complete DMEM medium.
2. Plates are treated as described above using LPS (10 ng/mL) with either zVAD.fmk (50 μM) or IDN-6556 (10 μM) +/−

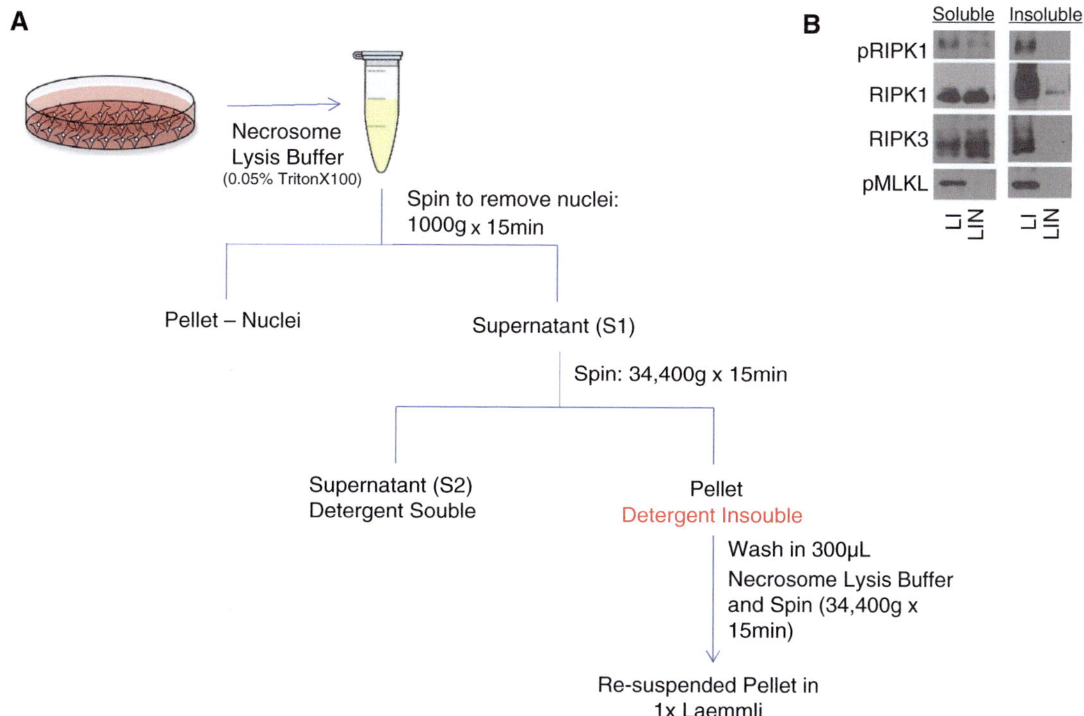

Fig. 3 Necrosome isolation by differential centrifugation. (**a**) Scheme of necrosome isolation. (**b**) Western analysis of pRIPK1, RIPK1, RIPK3, and pMLKL in detergent soluble and insoluble fractions from RAW264.7 cells treated for 3–4 h with LPS/IDN (LI) and LPS/IDN/Nec-1s (LIN). LPS = 10 ng/mL, IDN-6556 = 10 μM, and Nec-1s = 30 μM

Nec-1s (30 μM) to induce necrosome formation and necroptosis. Plates are maintained in an incubator at 37 °C with 5% CO_2 during the treatment period.

3. A minimal volume of lysate should be added to the top of a sucrose gradient prior to velocity sedimentation to maximize resolution. Accordingly, all the cells from the same treatment group are pooled together and lysed in the total volume of ≤350 μL. To achieve this, plates are placed on ice and washed with 2 mL cold PBS per plate. An additional 1.5 mL of cold PBS is added to each plate and adherent cells are collected using a cell scraper. Adherent cells are combined with media and PBS wash and centrifuged at 400 × *g* for 5 min at 4 °C.
4. Supernatants from the spin are aspirated and cell pellets are resuspended in 350 μL of Necrosome Lysis Buffer by pipetting (the lysate will be thick) (*see* **Note 6**).
5. Lysates are vortexed briefly (5–10 s, setting of 4.5) and then flash-frozen on dry ice.
6. Lysates are thawed at room temperature or on ice. Samples are vortexed for 5–10 s prior to centrifugation at 1000 × *g* for 15 min at 4 °C in a tabletop centrifuge to pellet nuclei.

7. Nuclear pellets are discarded and protein concentrations are normalized following protein assay. Typically, concentrations should be in 4–6 mg/mL range.
8. 300 μL of each sample is loaded onto a 10–50% linear sucrose gradient.
9. Gradients are subjected to velocity sedimentation using an ultracentrifuge at ~250,000 × g for 2.5–3 h at 4 °C.
10. Following sedimentation, 13–14 × 1 mL fractions are collected using a Gilson FC 203B fraction collector.
11. 120 μL of each sample can be diluted in 4× Laemmli sample buffer and boiled for 5–10 min to prepare samples for Western blot analysis. This is typically sufficient for 5–7 Western blots for RIPK1, RIPK3, MLKL, and corresponding phosphoproteins.
12. Using this method, coenrichment of RIPK1, RIPK3, and MLKL in LPS/zVAD or LPS/IDN samples is typically observed in higher density sucrose fractions (fraction numbers: 12–13) compared to samples treated with Nec-1s as well. If high sucrose concentration interferes with Western analysis or greater sensitivity of assay is desired, protein from 1 mL samples can be precipitated by chloroform–methanol extraction prior to preparation of sample for Western analysis (*see* **Note 13**). Examples of sucrose gradient fractions from RAW264.7 cells undergoing necroptosis blotted for necrosome components are shown in Fig. 4.

3.7 Immunoprecipitation of RIPK1 (Fig. 5)

Enrichment of RIPK1, RIPK3, MLKL, and their activated forms in detergent-insoluble cellular fractions correlates with the activation of RIPK1 kinase and induction of necroptosis [21]. Nevertheless, the approach does not demonstrate direct association

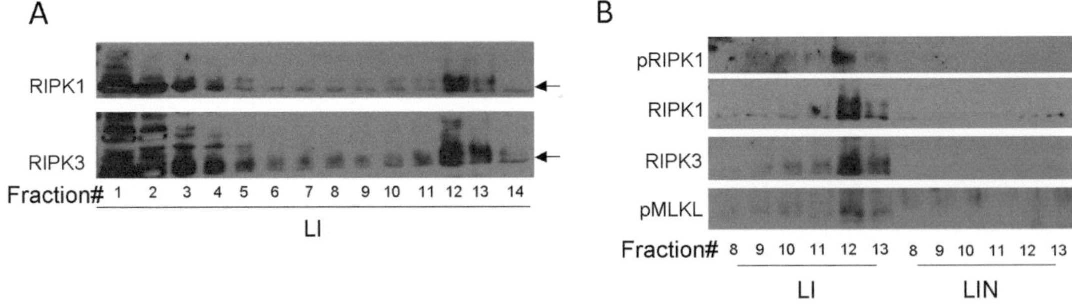

Fig. 4 Necrosome isolation by sucrose gradient and velocity sedimentation. (**a**) Western analysis of RIPK1 and RIPK3 in fractions 1–14 from RAW264.7 macrophages treated for 3–4 h with LPS/IDN (LI). Black arrows identify RIPK1 and RIPK3, respectively. (**b**) Western analysis of pRIPK1, RIPK1, RIPK3, and pMLKL in high-density sucrose fractions (#8–13) in RAW264.7 macrophages treated for 3–4 h with LPS/IDN (LI) or LPS/IDN/Nec-1s (LIN). LPS = 10 ng/mL, IDN-5665 = 10 μM, and/or Nec-1s = 30 μM

of necrosome components. This can be further addressed using coimmunoprecipitation.

1. Postnuclear RAW264.7 cell lysates are prepared as discussed in Subheading 3.5. However, lysates are not flash-frozen prior immunoprecipitation of RIPK1. In lieu, after homogenization, lysates are vortexed briefly (5–10 s at setting 4.5), kept on ice for 10 min, and then vortexed again prior to centrifugation at 1000 × *g* for 15 min.
2. Postnuclear lysates are diluted to 2 mg/mL. 45 μL of lysate is diluted in 4× Laemmli buffer as an 'Input' sample for Western analysis. This volume is sufficient for >10 blots of RIPK1, RIPK3, and MLKL.
3. Each lysate sample is divided into two equal aliquots (typically 250–300 μL). Anti-RIPK1 antibody is added to one of the samples at 1:100 dilution. Total rabbit IgG (Santa Cruz) can be added to the second sample, which serves as a negative (bead-only) control. Samples are rotated overnight at 4 °C.
4. The following morning, 10 μL of Protein A-conjugated magnetic beads per sample are washed 3× with 300 μL of Necrosome Lysis Buffer using a magnetic rack. After the final wash, beads are mixed with each IP samples and rotated additional 3–4 h at 4 °C.
5. After the incubation, beads are collected using a magnetic rack and boiled in 50 μL of 1× Laemmli sample buffer. Additional washes with water, PBS or lysis buffer can be included prior to elution to reduce background. Wash conditions need to be optimized for specific interacting factors. The sample is sufficient for >2–4 blots of RIPK1 and associated factors (*see* **Note 14**). Figure 5 shows a coimmunoprecipitation analysis showing an interaction between RIPK1 and RIPK3 in RAW264.7 cells treated with LPS + IDN-5665 to induce necroptosis.

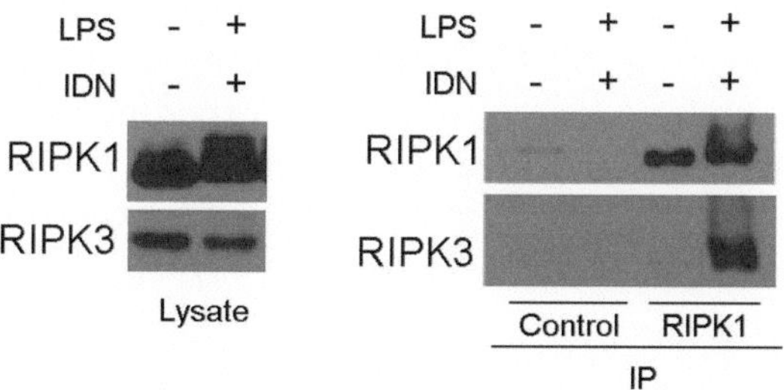

Fig. 5 Immunoprecipitation of RIPK1 and western analysis of RIPK1 and RIPK3 from RAW264.7 macrophages treated with LPS/IDN for 3–4 h. LPS = 10 ng/mL, IDN-5665 = 10 μM, and/or Nec-1s = 30 μM. Control = Beads only

4 Notes

1. Different subclones of widely used cell lines may display widely variable sensitivities to necroptosis. Sublines of mouse NIH3T3 fibroblasts have been reported to lose expression of RIPK3 and become resistant to necroptosis [24]. Similarly, we found that different MEF isolates display widely variable activation of necroptosis in response to TNF/zVAD. In our hands, A3 subclone of Jurkat cells displays higher sensitivity to TNF/CHX/zVAD compared to E6 Jurkat subclone. Different subclones of L929 cells may also display variable activation of necroptosis in response to zVAD or TNFα. Thus, specific concentrations of necroptosis inducers and timing of the viability assays need to be adjusted prior to performing screen.
2. While FADD-deficient Jurkat cells provide a very convenient system to identify necroptosis modulators, multiple other cellular systems can be used. For example, wild type Jurkat cells can be treated with 10 ng/mL human TNFα, 1 μg/mL CHX and 100 mM zVAD.fmk to achieve >60% cell death in 24 h. L929 cells can be treated with 10 ng/mL mouse TNFα or 20–50 μM zVAD.fmk for 16–24 h. Combination of TNF/zVAD will result in faster death within 12 h. RAW264.7 cells can be treated with 10–100 ng/mL LPS and 50 μM zVAD.fmk to undergo 50–70% death at 24 h. In all cases zVAD.fmk can be replaced by comparable concentrations of qVD-OPH or 10–20 μM IDN-6556. Addition of SM164 or 5z-7-oxozeaenol at 100–1000 nM may be used to achieve more robust activation of necroptosis in multiple cell types.
3. Because necrosomes are the likely downstream signaling effectors and their formation typically requires 1.5–3 h, it is possible to add TNFα to the culture medium before treating the cells, followed by the addition of compounds without significant loss in inhibition. However, this also depends on the point of activity of an inhibitor, and the specifics of the equipment used for the screening, e.g., total time required for cell plating and compound addition.
4. For isolation of necrosome components using differential centrifugation, experiments can be carried out on a smaller scale in 6-well plates for optimization purposes. In a 6-well plate format, cells can be seeded overnight at 2 million/well. In 6-well plate format, 150–200 μL Necrosome Lysis Buffer per well is appropriate for harvesting the cells.
5. The duration of treatment time may vary from one laboratory to another. Typically, reliable enrichment of RIPK1 and RIPK3 in detergent insoluble fractions/necrosome formation can be demonstrated when ~>20% of treated macrophages have begun to lose adherence to the plate.

6. Nonadherent cells in LPS/zVAD or LPS/IDN treated samples are recovered by collecting the treatment media and spinning down live cells/cellular debris at 400 × *g* for 5 min. (This can be done at room temperature or at 4 °C.) Supernatant may then be aspirated and cell pellet resuspended in the aliquot of lysis buffer used to harvest adherent cells from the corresponding treatment condition.
7. If starting with a confluent 10-cm plate (1 × 10^7 cells/plate per condition), 50 μL of S2 is sufficient for >15 blots for RIPK1, RIPK3, and MLKL and ~5 phosphoprotein blots.
8. To remove residual supernatant, avoid using vacuum-assisted aspiration as this may result in loss of the pellet. Using a pipetmen and gel loading tips, remove the supernatant, leaving behind less than 5–10 μL of fluid.
9. The whitish pellet should be visible at the bottom of the 1.7 mL Eppendorf tube after the spin. Pipetting the intact pellet 3–5 times in the new aliquot of lysis buffer is sufficient for the wash.
10. Remove as much supernatant as possible. This might be aided by the use of a gel-loading tip to remove small volumes above the pellet.
11. For samples generated from 6-well plates, we use 30–40 μL as opposed to 50–60 μL of 1× Laemmli sample loading buffer.
12. Sucrose gradients are prepared by adding 6.1 mL of 10% sucrose (with 3 μM DTT) to an ultracentrifuge tube followed by addition of 6.1 mL of 50% sucrose (with 3 μM DTT) using a syringe with cannula to displace lower percentage solution with the higher percentage solution. Overflow caps are added to the tubes and a Biocomp Gradient Master is used to generate a 10–50% linear sucrose gradient. Gradients are stored at 4 °C for 30 min before use.
13. Chloroform–methanol extraction serves to concentrate protein and reduce/eliminate sucrose, which may interfere with Western blotting. The procedure is carried out at room temperature in the fume hood. 150 μL of sucrose fraction is mixed with 600 μL of methanol. 150 μL of chloroform is added to each tube and samples are vortexed briefly at maximal speed. 450 μL of distilled water is added to each tube and samples are vortexed again. Samples are centrifuged at 14,000 × *g* for 5 min at 4 °C. The upper phase is carefully removed without disrupting the lower phase. 650 μL of methanol is added to the tube and the tube is inverted 3× times to mix. Samples are then spun at 34,400 × *g* for 5 min at 4 °C. Following the spin, a white pellet is observed at the bottom of the tube. The supernatant is removed without disrupting the pellet and the pellet is allowed to air-dry for 2–3 min. The pellet is then

resuspended in 15–20 μL of 4× Laemmli sample loading buffer. It is critical to resuspend the sample in 4× and not 1× loading buffer as residual methanol in the pellet may affect loading the samples onto the SDS-PAGE in 1× Laemmli sample loading buffer.

14. In examining proteins that have coimmunoprecipitated with RIPK1 by Western, we recommend using the EasyBlot kit (GeneTex GTX225856–01). These reagents serve to reduce background nonspecific binding caused by Protein A and IgG heavy chain.

References

1. Galluzzi L, Kepp O, Chan FK, Kroemer G (2016) Necroptosis: mechanisms and relevance to disease. Annu Rev Pathol. https://doi.org/10.1146/annurev-pathol-052016-100247
2. Zhao H, Jaffer T, Eguchi S, Wang Z, Linkermann A, Ma D (2015) Role of necroptosis in the pathogenesis of solid organ injury. Cell Death Dis 6:e1975. https://doi.org/10.1038/cddis.2015.316
3. Linkermann A, Green DR (2014) Necroptosis. N Engl J Med 370(5):455–465. https://doi.org/10.1056/NEJMra1310050
4. Degterev A, Linkermann A (2016) Generation of small molecules to interfere with regulated necrosis. Cell Mol Life Sci 73(11–12):2251–2267. https://doi.org/10.1007/s00018-016-2198-x
5. Sun L, Wang H, Wang Z, He S, Chen S, Liao D, Wang L, Yan J, Liu W, Lei X, Wang X (2012) Mixed lineage kinase domain-like protein mediates necrosis signaling downstream of RIP3 kinase. Cell 148(1–2):213–227. https://doi.org/10.1016/j.cell.2011.11.031
6. Degterev A, Huang Z, Boyce M, Li Y, Jagtap P, Mizushima N, Cuny GD, Mitchison TJ, Moskowitz MA, Yuan J (2005) Chemical inhibitor of nonapoptotic cell death with therapeutic potential for ischemic brain injury. Nat Chem Biol 1(2):112–119. https://doi.org/10.1038/nchembio711
7. Ren Y, Su Y, Sun L, He S, Meng L, Liao D, Liu X, Ma Y, Liu C, Li S, Ruan H, Lei X, Wang X, Zhang Z (2016) Discovery of a highly potent, selective, and metabolically stable inhibitor of receptor-interacting protein 1 (RIP1) for the treatment of systemic inflammatory response syndrome. J Med Chem. https://doi.org/10.1021/acs.jmedchem.6b01196
8. Harris PA, King BW, Bandyopadhyay D, Berger SB, Campobasso N, Capriotti CA, Cox JA, Dare L, Dong X, Finger JN, Grady LC, Hoffman SJ, Jeong JU, Kang J, Kasparcova V, Lakdawala AS, Lehr R, McNulty DE, Nagilla R, Ouellette MT, Pao CS, Rendina AR, Schaeffer MC, Summerfield JD, Swift BA, Totoritis RD, Ward P, Zhang A, Zhang D, Marquis RW, Bertin J, Gough PJ (2016) DNA-encoded library screening identifies Benzo[b][1,4]oxazepin-4-ones as highly potent and Monoselective receptor interacting protein 1 kinase inhibitors. J Med Chem 59(5):2163–2178. https://doi.org/10.1021/acs.jmedchem.5b01898
9. Berger SB, Harris P, Nagilla R, Kasparcova V, Hoffman S, Swift B, Dare L, Schaeffer M, Capriotti C, Ouellette M, King BW, Wisnoski D, Cox J, Reilly M, Marquis RW, Bertin J, Gough PJ (2015) Characterization of GSK'963: a structurally distinct, potent and selective inhibitor of RIP1 kinase. Cell Death Discov (1):15009. https://doi.org/10.1038/cddiscovery.2015.9
10. Harris PA, Bandyopadhyay D, Berger SB, Campobasso N, Capriotti CA, Cox JA, Dare L, Finger JN, Hoffman SJ, Kahler KM, Lehr R, Lich JD, Nagilla R, Nolte RT, Ouellette MT, Pao CS, Schaeffer MC, Smallwood A, Sun HH, Swift BA, Totoritis RD, Ward P, Marquis RW, Bertin J, Gough PJ (2013) Discovery of small molecule RIP1 kinase inhibitors for the treatment of pathologies associated with necroptosis. ACS Med Chem Lett 4(12):1238–1243. https://doi.org/10.1021/ml400382p
11. Mandal P, Berger SB, Pillay S, Moriwaki K, Huang C, Guo H, Lich JD, Finger J, Kasparcova V, Votta B, Ouellette M, King BW, Wisnoski D, Lakdawala AS, DeMartino MP, Casillas LN, Haile PA, Sehon CA, Marquis RW, Upton J, Daley-Bauer LP, Roback L, Ramia N, Dovey CM, Carette JE, Chan FK, Bertin J, Gough PJ, Mocarski ES, Kaiser WJ (2014) RIP3 induces apoptosis independent of pronecrotic kinase activity. Mol Cell

56(4):481–495. https://doi.org/10.1016/j.molcel.2014.10.021

12. Li D, Li C, Li L, Chen S, Wang L, Li Q, Wang X, Lei X, Shen Z (2016) Natural product Kongensin a is a non-canonical HSP90 inhibitor that blocks RIP3-dependent necroptosis. Cell Chem Biol 23(2):257–266. https://doi.org/10.1016/j.chembiol.2015.08.018

13. Fauster A, Rebsamen M, Huber KV, Bigenzahn JW, Stukalov A, Lardeau CH, Scorzoni S, Bruckner M, Gridling M, Parapatics K, Colinge J, Bennett KL, Kubicek S, Krautwald S, Linkermann A, Superti-Furga G (2015) A cellular screen identifies ponatinib and pazopanib as inhibitors of necroptosis. Cell Death Dis 6:e1767. https://doi.org/10.1038/cddis.2015.130

14. Najjar M, Suebsuwong C, Ray SS, Thapa RJ, Maki JL, Nogusa S, Shah S, Saleh D, Gough PJ, Bertin J, Yuan J, Balachandran S, Cuny GD, Degterev A (2015) Structure guided design of potent and selective ponatinib-based hybrid inhibitors for RIPK1. Cell Rep 10(11):1850–1860. https://doi.org/10.1016/j.celrep.2015.02.052

15. Li JX, Feng JM, Wang Y, Li XH, Chen XX, Su Y, Shen YY, Chen Y, Xiong B, Yang CH, Ding J, Miao ZH (2014) The B-Raf(V600E) inhibitor dabrafenib selectively inhibits RIP3 and alleviates acetaminophen-induced liver injury. Cell Death Dis 5:e1278. https://doi.org/10.1038/cddis.2014.241

16. Kaiser WJ, Sridharan H, Huang C, Mandal P, Upton JW, Gough PJ, Sehon CA, Marquis RW, Bertin J, Mocarski ES (2013) Toll-like receptor 3-mediated necrosis via TRIF, RIP3, and MLKL. J Biol Chem 288(43):31268–31279. https://doi.org/10.1074/jbc.M113.462341

17. Schworer SA, Smirnova II, Kurbatova I, Bagina U, Churova M, Fowler T, Roy AL, Degterev A, Poltorak A (2014) Toll-like receptor-mediated down-regulation of the deubiquitinase cylindromatosis (CYLD) protects macrophages from necroptosis in wild-derived mice. J Biol Chem 289(20):14422–14433. https://doi.org/10.1074/jbc.M114.547547

18. Dillon CP, Weinlich R, Rodriguez DA, Cripps JG, Quarato G, Gurung P, Verbist KC, Brewer TL, Llambi F, Gong YN, Janke LJ, Kelliher MA, Kanneganti TD, Green DR (2014) RIPK1 blocks early postnatal lethality mediated by caspase-8 and RIPK3. Cell 157(5):1189–1202. https://doi.org/10.1016/j.cell.2014.04.018

19. Upton JW, Kaiser WJ, Mocarski ES (2012) DAI/ZBP1/DLM-1 complexes with RIP3 to mediate virus-induced programmed necrosis that is targeted by murine cytomegalovirus vIRA. Cell Host Microbe 11(3):290–297. https://doi.org/10.1016/j.chom.2012.01.016

20. Li J, McQuade T, Siemer AB, Napetschnig J, Moriwaki K, Hsiao YS, Damko E, Moquin D, Walz T, McDermott A, Chan FK, Wu H (2012) The RIP1/RIP3 necrosome forms a functional amyloid signaling complex required for programmed necrosis. Cell 150(2):339–350. https://doi.org/10.1016/j.cell.2012.06.019

21. Moquin DM, McQuade T, Chan FK (2013) CYLD deubiquitinates RIP1 in the TNFalpha-induced necrosome to facilitate kinase activation and programmed necrosis. PLoS One 8(10):e76841. https://doi.org/10.1371/journal.pone.0076841

22. Najjar M, Saleh D, Zelic M, Nogusa S, Shah S, Tai A, Finger JN, Polykratis A, Gough PJ, Bertin J, Whalen MJ, Pasparakis M, Balachandran S, Kelliher M, Poltorak A, Degterev A (2016) RIPK1 and RIPK3 kinases promote cell-death-independent inflammation by toll-like receptor 4. Immunity 45(1):46–59. https://doi.org/10.1016/j.immuni.2016.06.007

23. Ofengeim D, Ito Y, Najafov A, Zhang Y, Shan B, DeWitt JP, Ye J, Zhang X, Chang A, Vakifahmetoglu-Norberg H, Geng J, Py B, Zhou W, Amin P, Berlink Lima J, Qi C, Yu Q, Trapp B, Yuan J (2015) Activation of necroptosis in multiple sclerosis. Cell Rep 10(11):1836–1849. https://doi.org/10.1016/j.celrep.2015.02.051

24. Zhang DW, Shao J, Lin J, Zhang N, Lu BJ, Lin SC, Dong MQ, Han J (2009) RIP3, an energy metabolism regulator that switches TNF-induced cell death from apoptosis to necrosis. Science 325(5938):332–336. https://doi.org/10.1126/science.1172308

Chapter 4

Distinguishing Necroptosis from Apoptosis

Inbar Shlomovitz, Sefi Zargarian, Ziv Erlich, Liat Edry-Botzer, and Motti Gerlic

Abstract

Apoptosis was the first programmed cell death to be defined—highly regulated and immunologically silent, as apoptotic bodies are being removed without triggering inflammation. Few decades later, necroptosis was discovered—uniquely regulated but inflammatory. As these two programmed cell death pathways may be initiated via similar pathways (death receptors and intracellular receptors) while being differently regulated and resulting in distinctive physiological consequences, the need for distinguishing apoptosis from necroptosis is required. Here we describe a series of distinguishing assays that use apoptotic- and necroptotic-distinct response to pharmacological interventions with specific death inhibitors, morphology and death-specific proteins involvement. The procedure includes cell death kinetics assessment and morphology monitoring of stimulated and pharmacologically treated-cells using flow cytometry and live imaging, with the detection of death-specific proteins using Immunoblot. The procedure described here is simple and thus can be adjusted to various experimental systems, enabling apoptosis to be distinguished from necroptosis in one's system of interest, without the need for more complex reagents such as genetic knockout models.

Key words Cell death, Apoptosis, Necroptosis, Caspase 3, MLKL

1 Introduction

Cell death was originally divided into two well-distinguished forms. Apoptosis was considered as a programmed and highly regulated cell death, characterized by the formation of apoptotic bodies that are removed without triggering inflammation [1, 2]. In contrast, necrosis was considered as an accidental trauma, an induced and unregulated cell death leading to inflammatory response as the cellular contents and debris are released into the tissue [3]. However, research in the cell death field during the last few decades has shed light on few unique regulated but inflammatory cell death pathways including necroptosis—a RIPK3/MLKL-dependent caspase-independent cell death [4–6]. As opposed to apoptosis, necroptosis is characterized by cell swelling, membrane permeabilization, and leakage of cytoplasmic content, resulting in the release of danger associated molecular patterns (DAMPs) and inflammation [7–9].

Adrian T. Ting (ed.), *Programmed Necrosis: Methods and Protocols*, Methods in Molecular Biology, vol. 1857, https://doi.org/10.1007/978-1-4939-8754-2_4,

Hence, distinguishing apoptosis from necroptosis in one's experimental system can be highly important, indicating distinctive immunological implications and different genetic and pharmacological intervention opportunities.

When death is triggered extrinsically, the cell's default fate will be apoptosis. In this setting, the initiator caspases, such as caspase 8, cleave and activate effector caspases, such as caspase 3, resulting in the final execution of apoptosis [10]. Simultaneously, caspase 8 cleaves and inactivates RIPK3, to prevent necroptosis [5, 11, 12]. When caspase 8 is blocked by different viral or bacterial pathogens or by pharmacological interventions, necroptosis might take place as an alternative pathway. In these settings, death signals trigger the dimerization and phosphorylation of RIPK1 and RIPK3 leading to the phosphorylation and aggregation of MLKL by RIPK3 [13–16]. The phosphorylated MLKL (pMLKL) then translocates to the plasma cell membrane, leading to necroptosis [17–19].

Although in most scenarios RIPK1 kinase activity is essential for the induction of necroptosis [20], RIPK1-independent forms of necroptosis are well known, enabling the acknowledgment of necroptosis as RIPK3/MLKL-dependent [11, 12, 21]. Thus, necroptosis can be prevented by genetic depletion of RIPK3 and MLKL, and pharmacologically inhibited by blocking RIPK1 kinase activity by necrostatin-1 (nec-1 and nec-1s) [20], blocking RIPK3 kinase activity by various inhibitors [22] and by preventing pMLKL membrane translocation by necrosulfonamide (NSA, available for human cells only) [16]. Hence, these manipulations, together with the characteristic morphology and cell signaling of apoptosis versus necroptosis, are used to distinguish one pathway from the other.

Since the use of genetic manipulations is usually not feasible for every laboratory and is more time consuming, a simple method to distinguish one pathway from the other is needed. Thus, assessing the effect of pancaspase inhibitors, together with the necroptosis specific inhibitors for RIPK1, RIPK3, and MLKL on cell death kinetics using flow cytometry and live imaging can serve as indication for the cell death pathway specificity. Live imaging also gives the added value of monitoring the morphological features of the dying cells in one's experimental system. The working assumption is that if cell death is completely blocked by the addition of pancaspase inhibitors, the induced death in one's system is caspase-dependent and therefore not necroptotic.

Following kinetics assessment, it is essential to validate death specificity by the detection of death-specific proteins using western blot in both lysates and supernatants [23]. The presence of cleaved caspases including caspase 3 indicate apoptosis, while detecting pMLKL, independently of total MLKL protein levels detected, is currently the most specific and sensitive indicator of necroptosis. The detection of pMLKL is known to be challenging. Importantly, we recently found and established pMLKL detection method in the supernatant, making necroptosis detection much easier [23]. Using this combination of several methods could compensate for not using genetic models and provide a simple and achievable distinguishing assay.

The procedure described here in detail refers to a given extracellular death signal, TNFα with SMAC mimetics, in known cell lines used in our lab and established in the field [24]. However, it can be easily modified to one's experimental system using one's death-inducing treatments and cells of interest. Additionally, it can even be manipulated for distinguishing cell death type in in-vivo samples.

2 Materials

2.1 Inducing Death in the Presence of Specific Death Inhibitors

1. Recombinant human TNFα, CAS #94948-59-1.
2. The SMAC mimetic AZD5582 dihydrochloride, CAS #1258392-53-8.
3. Pancaspase inhibitor Z-VAD-fmk, CAS #187389-52-2 (*see* **Note 1**).
4. Necrostatin-1S (nec-1s), CAS #852391-15-2.
5. GSK'872, CAS #1346546-69-7.
6. Necrosulfonamide (NSA), CAS #1360614-48-7.
7. U937 cell culture medium: RPMI 1640 supplemented with 10% heat-inactivated FBS (unless mentioned otherwise), 100 units/mL penicillin G sodium salt, 100 μg/mL streptomycin sulfate, and 10 mM HEPES buffer solution.
8. HaCaT and L929 cell culture medium: DMEM (Dulbecco's Modified Eagle Medium) medium, supplemented with 10% heat-inactivated FBS (unless mentioned otherwise), 100 units/mL penicillin G sodium salt, 100 μg/mL streptomycin sulfate, and 10 mM HEPES buffer solution.

2.2 Assessing Cell Death Kinetics

1. Propidium iodide (PI) solution, CAS #25535-16-4. Dissolve 1 mg in 1 mL of cell culture grade water for stock solution. Store in aliquots of 500 μL at 4 °C protected from light.
2. Annexin-V, CAS #136107-94-3. We use annexin-V conjugated to APC when NSA is being used and to FITC while it is not used (*see* **Note 8**).
3. Annexin-V Binding buffer: 100 mM HEPES, 140 mM NaCl, 25 mM $CaCl_2$, pH 7.4. (Commercially available as 5× or 10×, should be diluted in deionized water before use and stored in 4 °C).
4. Live imaging apparatus. We use the EVOS FL Auto Imaging System and EVOS® Onstage Incubator (ThermoFisher Scientific) or IncuCyteZoom apparatus and perform data analysis using EVOS FL Auto software or IncuCyteZoom2016B software, respectively.

5. Flow cytometer. We use the 4-lasers 14 colors Attune NxT instrument (ThermoFisher Scientific). Results are analyzed using FlowJo V10.2 software.

2.3 Detecting Death-Specific Protein Involvement by Immunoblot

2.3.1 Collecting Samples

1. RIPA lysis buffer. Adjust 1 M Tris buffer to pH 7.5 with HCL. Add to a final concentration 1% (v/v) Triton X-100, 1% (w/v) sodium deoxycholate, 150 mM NaCl and 0.1% (w/v) SDS. Store at 4 °C.
2. Halt protease and the phosphatase inhibitor single-use cocktail, EDTA-free (ThermoFisher Scientific) containing aprotinin, bestatin, E-64, leupeptin, sodium fluoride, sodium orthovanadate, sodium pyrophosphate, and β-glycerophosphate. Should be added to RIPA lysis buffer in 1:100 dilution immediately prior to use.

2.3.2 Preparing Lysate Samples

1. 5× sample buffer. Add 80 mg bromophenol blue to 5 mL of 1.5 M Tris–HCl pH 6.8. Add 10 mL of glycerol and mix. Add 2 g of SDS and mix (SDS will take few minutes to dissolve). Add 5 mL of β-mercaptoethanol and mix. Store in aliquots of 500 μL at −80 °C.

2.3.3 Preparing Supernatant Samples

1. Methanol, CAS #67-56-1.
2. Chloroform CAS #67-66-3.

2.3.4 SDS-PAGE and Immunoblotting

1. Polyacrylamide gel. We use either 10% or 4–20% precast polyacrylamide gels.
2. 10× running buffer: 250 mM Tris base, 1.92 M glycine, 1% SDS, pH 8.3.
3. Transfer apparatus. We use Trans-Blot Turbo transfer system (BIORAD) using the matching nitrocellulose midi transfer packs.
4. TBST: 50 mM Tris–HCl pH 7.5, 150 mM NaCl (TBS), 0.05% Tween 20.
5. Blocking buffer: 50 mM Tris–HCl pH 7.5, 150 mM NaCl (TBS), 0.05% Tween 20, 5% (w/v) skim milk powder.
6. Antibodies. Table 1 lists antibodies that are used for immunoblotting.

Table 1
Antibodies

		Dilution	Source
Primary antibody	Cleaved caspase-3	1:1000	9961, Cell Signaling
	Human pMLKL	1:1000	187,091, Abcam
	Mouse pMLKL	1:1000	196,436, Abcam
	Total MLKL	1:1000	MABC604, Merck Millipore
Secondary antibody (HRP conjugated)	Donkey anti-rabbit	1:5000	Jackson ImmunoResearch Labs
	Goat anti-rat	1:5000	

7. Detection instrument. Blots are read in LI-COR Odyssey Fc and analyzed in Image Studio version 5.0 software.
8. Stripping buffer. We use the commercial stripping buffer NewBlot IR stripping buffer (5×) diluted to 1× in distilled water.

3 Methods

All cells are cultured in incubators at 37 °C, 5% CO_2, and 95% RH. All methods are detailed descriptively for the human monocytic suspension U937 cell line and briefly for adherent cells in case an adjustment is needed.

3.1 Inducing Death in the Presence of Specific Death Inhibitors

The use of chemical components may cause cell toxicity from certain dose [22]. As components toxicity may vary between different cells, calibrating cell toxicity in respond to Z-VAD-fmk, nec-1s, GSK'872, and NSA (*see* **Note 2**) is necessary in establishing this assay.

3.1.1 Calibrating Inhibitors Toxicity

1. Suspend 1.2×10^7 U937 cells in 6 mL of serum-free cell culture medium (RPMI 1640) containing 2 μg/mL PI (1:500 dilution from stock) in sterile 96-well plate. In case of adherent cells, plate the cells for overnight incubation prior to the day of the experiment in 10% serum-cell culture medium.
2. Seed 100 μL (2×10^5 cells) per well in 51 wells of sterile 96-well plate.
3. Prepare 350 μL of at least four doses of Z-VAD-fmk (pancaspase inhibitor) ranging between 20–80 μM (10–40 μg/mL) (will be further diluted by twofold) in serum-free RPMI 1640. (In case of adherent cells *see* **Note 3**)
4. Prepare 350 μL of at least four doses of nec-1s (RIPK1 inhibitor) ranging between 3.6–14.4 μM (1–4 μg/mL) (will be further diluted by twofold) in serum-free RPMI 1640. (In case of adherent cells *see* **Note 3**).
5. Prepare 350 μL of at least four doses of GSK'872 (RIPK3 inhibitor) ranging between 5.2–15.6 μM (2–6 μg/mL) (will be further diluted by twofold) in serum-free RPMI 1640. This inhibitor is specifically known to induce apoptosis in some cells [25] (*see* Fig. 2c). (In case of adherent cells *see* **Note 3**.)
6. Prepare 350 μL of at least four doses of NSA (MLKL inhibitor) ranging between 1–4 μM (0.45–1.8 μg/mL) (will be further diluted by twofold) in serum-free RPMI 1640. (In case of adherent cells *see* **Note 3**.)
7. Prepare 500 μL of serum-free RPMI 1640 to serve as untreated control group.
8. Transfer 100 μL of each treatment to cells in triplicate. In case of adherent cells, discard media of overnight incubation and add fresh medium before adding the treatments.

9. Record plate in live Imaging for 24 h (recording specific frequency in not critical in this calibration step, we usually use 15 min intervals).
10. Analyze results and determine the highest doses of inhibitors that cause minimal cell toxicity.

3.1.2 Inducing Death in Pharmacologically Treated Cells

1. Seed 2 × 10^5 U937 cells suspended in 100 μL serum-free cell culture medium (RPMI 1640) in a sterile 96-well plate. In case of adherent cells, plate the 3–5 × 10^5 cells per well for overnight incubation prior to the day of the experiment in 10% serum-cell culture medium.
2. Prepare 350 μL of the following samples in serum-free cell culture medium (RPMI 1640), unless mentioned otherwise (note that the inhibitors concentrations are taken from Subheading 3.1.1):
   (a) Cell culture medium (RPMI 1640) only—prepare 650 μL.
   (b) Pancaspase inhibitor Z-VAD-fmk (40 μM, 20 μg/mL) to test for apoptosis.
   (c) RIPKl inhibitor nec-1s (10 μm, 2.8 μg/mL)—inhibitor of necroptosis.
   (d) RIPK3 inhibitor GSK'872 (10.4 μM, 4 μg/mL)—inhibitor of necroptosis.
   (e) MLKL inhibitor NSA (2 μM, 0.9 μg/mL)—inhibitor of necroptosis.
   (f) Z-VAD-fmk + nec-1s.
   (g) Z-VAD-fmk + GSK'872.
   (h) Z-VAD-fmk + NSA.

   Notice that our inhibitors final concentration after adding them to the cells will be: Z-VAD-fmk (20 μM, 10 μg/mL), nec-1s (5 μm, 1.4 μg/mL), GSK'872 (5.2 μM, 2 μg/mL), NSA (1 μM, 0.45 μg/mL). (In case of adherent cells *see* **Note 3**.)

   In case preparing the plate for the subsequent live imaging procedure (Subheading 3.2.1.), the medium should contain 2 μg/mL PI and ~25 ng/mL annexin-V-FITC (the concentration is LOT specific) (*see* **Note 8**).
3. Pretreat cells with the various inhibitors combinations by adding triplicates of 100 μL from each sample *a to h*.
4. Incubate for 30 min in 37 °C.
5. Prepare concentrated death inducer of interest, so only up to 10 μL should be added per well to induce death.
6. Transfer to cells in triplicate (note to leave one untreated triplicate as control). We use TNFα (1.15 nM, 20 ng/mL) + the SMAC mimetic AZD5582 [10 μM, 10 μg/mL], therefore:

Prepare 60 μL of TNFα (115 nM, 2 μg/mL) and add 2 μL per well.

Prepare 60 μL of AZD5582 (1 mM, 1 mg/mL) and add 2 μL per well.

3.2 Assessing Cell Death Kinetics

The core concept of cell death real-time assessment lies in the staining of cells using propidium Iodide (PI) and annexin-V. PI is a cell-impermeable fluorescent dye that binds nucleic acids only once plasma and nuclear membrane integrity is damaged. Annexin-V, which is also cell-impermeable, binds to phosphatidylserine at early stages of cell death (apoptosis and necroptosis) upon phosphatidylserine flipping to the outer plasma membrane (*see* **Note 4**).

3.2.1 Qualitative and Morphological Monitoring by Live Imaging

1. Set your microscope for taking images of each well of interest every few minutes as frequently as needed and if possible with objective lens of ×10/×20 in bright field, Texas Red, and GFP light cubes (*see* **Notes 5–8**). For example, U937 images are taken every 15 min for 6 h.
2. Gather the images taken in each time point of each well to create a movie.
3. Examine qualitatively.

   If Z-VAD-fmk treated death-induced cells seem similar to the untreated ones (e.g. cell death was inhibited), apoptosis is probably the dominant death pathway in one's system. If not, the option that necroptosis is occurring should be further tested. If the addition of necroptosis-inhibitors to untreated (*see* Fig. 2c) and Z-VAD-fmk (*see* Fig. 2d) treated death-induced cells inhibited death, it is highly suggestive that necroptosis is occuring (*see* **Note 9** and Fig. 1a).
4. Examine the morphological features of treated cells in each one of the relevant fields and characterize accordingly. Table 2 shows some of the morphological characteristics of the two forms of cell death:

3.2.2 Quantitative Assessment by Flow Cytometry

This procedure should also be done following Subheading 3.1.2, independently of the above live-imaging procedure.

1. In each time point, transfer 30 μL of U937 cells to a clean U-bottom 96-well plate. Usually we monitor death kinetic every 30 min. For adherent cells, each time point needs to have separated replicate wells (*see* **Note 10**).
2. Centrifuge U-bottom plate at 400 × *g* for 5 min and discard the supernatant.
3. Resuspend the cell pellets with 50 μL of annexin-V-APC (this is important as we notice that NSA have an autofluorescent in the flow cytometry 488 laser, (*see* **Note 8**)), diluted 1:500 in binding buffer. For adherent cells, gently trypsinize and wash

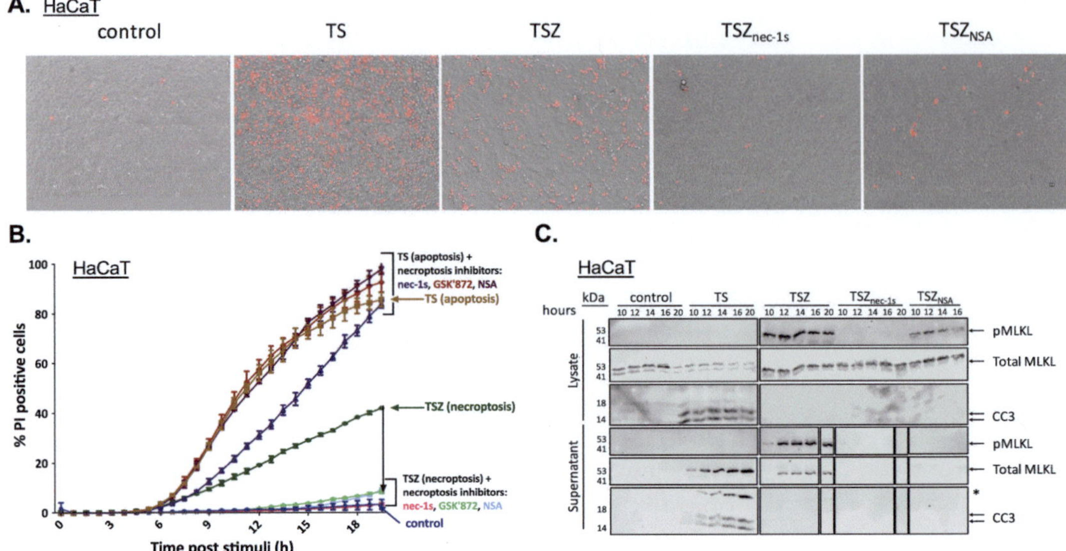

Fig. 1 Characterizing death in HaCaT cells. HaCaT cells were stimulated for either apoptosis (TS), necroptosis (TSZ) or left untreated (control) with and without the addition on nec-1s, GSK'872 and NSA as describes in Subheading 3.1.2. (**a**) 14 h post stimulation, PI was added to the medium and images were taken using live microscopy for qualitative assessment as described in Subheading 3.2.1. (**b**) 30,000 HaCat cells per well in 96-well plate were seeded and treated. The plate was placed in IncuCyteZOOM apparatus and two images per well were recorded in indicated time points. Death kinetics data was quantitatively analyzed using IncuCyteZoom2016B software. (**c**) 10^6 cells were harvested and supernatant was collected at indicated time points post stimulation, and the death-specific proteins pMLKL and cleaved caspase 3 (CC3) and total MLKL as control were detected using Western Blot as described in Subheading 3.3

Table 2
Characteristic morphological features

	Apoptosis	Necroptosis
Bright field	Reduction of cellular volume—shrinking cells	Increasing cellular volume—swelling cells
	Big membrane blebbing	Permeabilization of the plasma membrane
PI	Small dots will appear in the cells ➔ reflects nuclear fragmentation or condensation.	Wide round "cytoplasmic-like" spread uptake ➔ reflects nuclear decondensation.
Annexin-V	Plasma membrane spread	Plasma membrane spread

cells once with serum-containing media before incubation with annexin-V.

4. Incubate plate for 15 min at room temperature protected from light.
5. Add 200 μL of DMEM medium supplemented with PI for final concentration (in total volume) of 1 μg/mL per well (*see* **Note 11**).

Table 3
Flow cytometry analysis strategy

Cells status	Population
Viable	Double negative (annexin-V^-PI^-)
Early apoptosis/necroptosis	Single positive for annexin-V (annexin-V^+PI^-) (*see* **Note 11**)
Late apoptosis/necroptosis	Double positive (Annexin-V^+PI^+)

6. Analyze samples by a flow cytometer according to the dyes' excitation/emission maxima:
   PI 493/636 nm.
   Annexin-APC 650/660 nm.
7. Analysis of the results will distinguish between three major different populations. The strategy for distinguishing the three populations is shown in Table 3.

   In U937 cells, early necroptosis is detected around 2.5 h post necroptosis stimulation and early apoptosis is detected 3.5 h after apoptosis stimulation.
8. Examine quantitatively by creating graph presenting % annexin-V^+PI^- and annexin-V^+PI^+ cells in each time point for every treatment combination (*see* Fig. 2a and **Note 12**).

   If Z-VAD-fmk treated death-induced cells seem similar to the untreated ones (e.g. cell death was inhibited), apoptosis is probably the dominant death pathway in one's system. If not, the option that necroptosis is occurring should be further tested. If the addition of necroptosis-inhibitors to untreated (*see* Fig. 2c) and Z-VAD-fmk (*see* Fig. 2d) treated death-induced cells inhibited death it is highly suggestive that necroptosis is occuring (*see* **Note 9** and Fig. 2a).

3.3 Detecting Death-Specific Protein Involvement by Immunoblot

After cell death kinetics have indicated the presence of apoptosis versus necroptosis, further validation is necessary. Treat and induce your cells as in Subheading 3.1.2 using higher number of cells (for U937 cells, seed at least 5×10^5 cells per well for 100 μL lysate sample) (*see* **Note 13**). Based on data obtained using the previous procedures discussed, an appropriate time point for sample collection should be chosen. Generally, 50% PI positive cell population is enough for the detection of the specific-death markers: cleaved caspase-3 and pMLKL in apoptotic and necroptotic cells, respectively. Choosing earlier time point, with higher presence of annexin-V single positive population could enable better detection of pMLKL in cells lysate, while it could be challenging later [23].

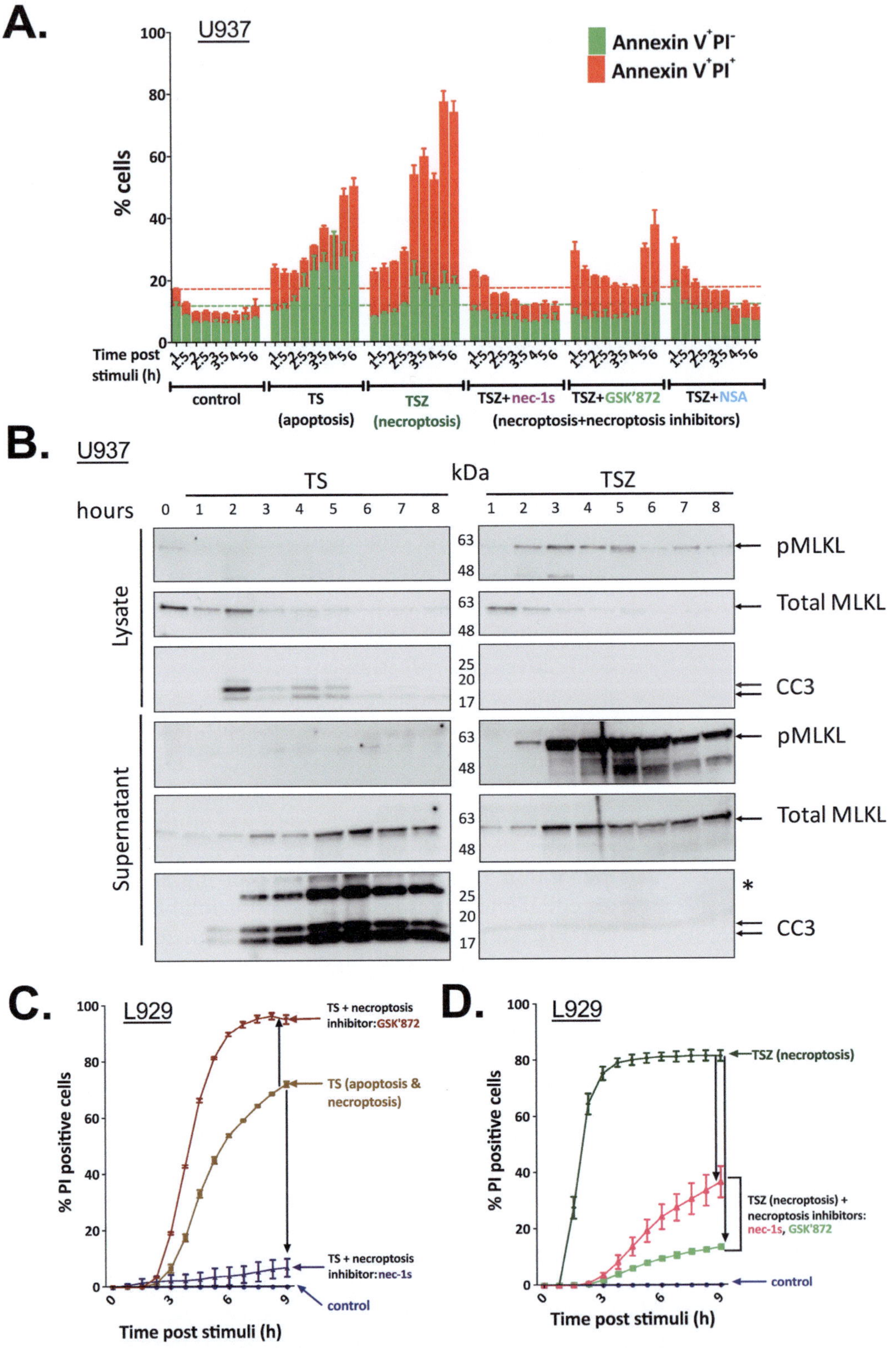

Fig. 2 Characterizing death in U937 and L929 cells. (**a, b**) U937 cells were stimulated for either apoptosis (TS), necroptosis (TSZ) or left untreated (None) as describes in Subheading 3.1.2. (**a**) Cell viability was measured at indicated time points post stimulation using annexin-V-FITC/PI staining and then quantitatively analyzed by

3.3.1 Collecting Samples from Suspension Cells

1. Transfer suspension cells and their supernatant to a suitable centrifuge tube (*see* **Note 13 and 14**).
2. Centrifuge tube for 5 min at 400 × *g* at 4 °C. (*see* **Notes 15** and **16**).
3. Transfer the upper liquid phase to a new tube and keep on ice. This is your supernatant.
4. Resuspend pellet in 100 μL RIPA buffer (supplemented with 1× protease and phosphatase inhibitors) (*see* **Note 17**).

3.3.2 Collecting Samples from Adherent Cells

1. Collect the media from the plate to a suitable tube. Centrifuge the tube for 5 min at 400 × *g* at 4 °C (*see* **Note 16**).
2. Transfer the upper liquid phase to a new tube and keep on ice. This is your supernatant. Keep the pellets on ice for later. These pellets contain broken and detached cells.
3. Add enough trypsin to cover the surface of the plate, incubate at 37 °C for few minutes.
4. Add 10% serum-containing cell culture medium to trypsinized cells and combine with pellets from **step 2**.
5. Centrifuge the tube for 5 min at 400 × *g* at 4 °C.
6. Discard the liquid phase. Resuspend pellet in 100 μL RIPA buffer (supplemented with 1× protease and phosphatase inhibitors) (*see* **Note 17**).

3.3.3 Preparing Lysate Samples

1. Incubate RIPA-containing samples for 15 min on ice (*see* **Note 18**).
2. Centrifuge samples for 15 min at 15,000 × *g* at 4 °C.
3. Collect the liquid phase to a new tube. This is your lysate.
4. Add 5× sample buffer to the desired lysate volume for final concentration of 1× (dilute 1:5) and incubate for 5 min at 95 °C. Then spin down.
5. The samples are now ready for polyacrylamide gel electrophoresis (PAGE). Samples can be stored at −20 °C until loading.

Fig. 2 (continued) flow cytometry as described in Subheading 3.2.1. (**b**) 10^6 cells were harvested and supernatant was collected at indicated time points post stimulation. The death-specific proteins pMLKL and cleaved caspase 3 (CC3) and total MLKL as controls were detected using western blotting as described in Subheading 3.3. (**c**, **d**) L929 cells were stimulated for necroptosis using either (TS) alone (**c**) or with the addition of pancaspase inhibitor Z-VAD-fmk (TSZ) (**d**) or left untreated (control) with and without the addition of nec-1s and GSK'872 as described in Subheading 3.1.2. NSA was not added to this murine cell line (*see* **Note 2**). Cells were seeded at 15,000 cells per well in a 96-well plate and treated. The plate was placed in an IncuCyteZOOM apparatus and two images per well were recorded at indicated time points. Death kinetics data was quantitatively analyzed using IncuCyteZoom2016B software

3.3.4 Preparing Supernatant Samples (Methanol-Chloroform)

1. Transfer 500 μL of your supernatant to a new 1.5 mL tube (*see* **Note 19**).
2. Add 500 μL methanol and 125 μL chloroform (for every 100 μL supernatant: 100 μL methanol +25 μL chloroform).
3. Vortex for 5 s.
4. Centrifuge for 5 min at 15,000 × *g* at 4 °C.
5. Using a pipette, gently discard the upper phase (above the white "ring" that separates the two phases) (*see* **Note 20**).
6. Add 500 μL methanol.
7. Vortex for 5 s.
8. Centrifuge for 5 min at 15,000 × *g* at 4 °C.
9. Using a pipette, gently discard the upper phase (methanol) as much as possible without disturbing the pellet.
10. Incubate open tubes for 5 min at 50 °C for methanol to evaporate. Further evaporation can be performed by leaving the tubes open in a fume-hood (*see* **Note 21**).
11. Add 5× sample buffer to the desired supernatant volume for final concentration of 1× (dilute 1:5) and incubate for 5 min at 95 °C. Then spin down.
12. The sample is now ready for polyacrylamide gel electrophoresis (PAGE). Samples can be stored at −20 °C until loading.

3.3.5 SDS–PAGE and Transfer

1. Load samples in 10% or 4–20% SDS-polyacrylamide gel so as all samples are obtained from the same cell number (equivalent to 1.5–2 × 10^5 cells for lysates and 3 × 10^5 cells for culture supernatants).
2. Run electrophoresis at 75 V until the samples have passed the stacking gel. You can increase the voltage up to 130 V once the samples have gone into the separating gel.
3. Once the 10 kDa ladder-band gets to the bottom of the gel, you should stop and turn off the electrophoresis (*see* **Note 22**).
4. Cut and discard the stacking gel.
5. Transfer to nitrocellulose as established in your lab.
6. Prepare blocking buffer in sufficient volume to cover your membrane (*see* **Note 23**).
7. Add the blocking buffer to the membrane and incubate on a shaker for at least 1 h at room temperature.

3.3.6 Immunoblotting

The recommended order for Immunoblotting the membrane is: (1) Anti-pMLKL antibody; (2) Anti-cleaved caspase 3 antibody; and (3) reprobe with anti-total MLKL after stripping (to prevent interruption by anti pMLKL antibodies). As mentioned in the introduction, the presence of cleaved caspase 3 indicate apoptosis, while detecting pMLKL is currently the most specific and sensitive indicator of necroptosis. To exclude the possibility that pMLKL was not detected due to protein detection limit, one should reprobe the membrane with anti-total MLKL antibody. In general, we anticipate that when necroptosis is induced, most of the MLKL will be found in its phosphorylated state (*see* Figs. 1c and 2b).

1. Add primary antibody in TBS with 0.05% Tween 20 (TBST) with 5% milk and 0.02% sodium azide and incubate overnight by shaking at 4 °C.
2. Wash the membrane six times by shaking with TBST for 5 min each at room temperature.
3. Add secondary antibody (also in TBST) and incubate for 1 h by shaking at room temperature.
4. Repeat wash as in **step 3**.
5. Add ECL (in case of using HRP antibody) for 5 min in the dark.
6. Dry out the ECL and read using your laboratory's preferred method.
7. Wash briefly in TBST and reprobe with the following primary antibody (unless stripping is required).
8. For stripping the blot, we use a commercial stripping buffer to remove primary and secondary antibodies as per manufacturer's instructions. (*see* **Note 24**).

4 Notes

1. We recommend avoiding using the pancaspase inhibitor Q-VD instead of Z-VAD-fmk since from our experimental experience Caspase 1 is not being efficiently blocked and pyroptosis can also be initiated in these settings [26]. If this is a concern, you may reprobe your membrane with anti-caspase 1 antibody.
2. Necrosulfonamide (NSA) efficiently inhibits necroptosis only in human cells, but not murine cells [16].
3. In case of adherent cells, always prepare inhibitors in the final concentration wanted in serum-free cell culture medium (*see* also **Note 19**) containing 1 μg/mL PI (1:1000 dilution from stock) if preparing for live imaging, as further dilution will not occur.
4. In opposition to the common assumption, we and others found that necroptotic cells expose phosphatidylserine to the outer plasma membrane after pMLKL translocation to the membrane. Thus, single annexin-V positive cells are not apoptosis-specific population and this population in necroptotic cells can indicate initiation of pMLKL detection in cells lysates by western blot [23]. We also found that using commercially available cell-impermeant amine-reactive dyes (e.g., Zombie dyes (Biolegend) or LiveDead dyes (Thermo

Fisher Scientific)), instead of PI, will detect earlier stages of plasma membranes integrity damage and will help to define a better time point for pMLKL detection in the lysate [23].

5. Live imaging by fluorescence microscopy can be toxic for live cells and can affect their viability. Different cell types will show different sensitivity to this. Sometimes long time-lapse may not be feasible despite the use of an incubator [27].
6. The death kinetics may be varied between different cell types. Thus, the frequency of image acquisition should be adjusted to the specific cell type. Working with a new cell type will often require some calibration.
7. Necroptosis and apoptosis processing may proceed differently in the same cell type. For that reason, various time points should be examined to assess death specificity.
8. NSA treatment causes high autofluorescence in flow cytometer 488 nm laser or GFP light cube of fluorescent microscope. Therefore, when using NSA, one should consider the use of other conjugated annexin-V other than annexin-V-FITC. We use annexin-V-APC.
9. Some cell lines do not undergo necroptotic death despite stimulation with necroptosis inducers. It might be due to loss of RIPK3 expression among many other possible changes that occur through cell culturing over the years.
10. Flow cytometry analysis of necroptotic adherent cells can be challenging as the dying cells get ruptured and totally broken. To analyze all cells when you detach them, make sure to take supernatant (which contains many detached cells), centrifuge at 400 × *g* for 5 min at room temperature and then combine trypsinzed cells with the pellets after thorough pipetting of trypsin-treated wells. In the event that percentage of annexin-V and PI positive cells do not seem to correlate with the death status qualitatively obtained by live imaging, we recommend comparing the live cell population (gated by the FSC versus SSC flow cytometry plots) between untreated sample against those treated with death inducers.
11. Annexin-V binding to phosphatidylserine is calcium-dependent. Calcium levels in different culture media can be variable. This could affect the annexin-V staining and therefore should be taken into account. While RPMI 1640 contains only 0.42 mM calcium, DMEM contains 1.8 mM and hence was chosen to be the suspending solution in this procedure.
12. It seems like annexin-V binding by necroptotic cells is higher than apoptotic ones. This observation is consistent in our lab with several cell types and cell death inducers.

13. For good death-specific protein detection by immunoblotting, lysates from 1.5–2 × 10^5 cells and supernatant from 3 × 10^5 cells should be loaded. Therefore, cell seeding should be calculated accordingly. Repeat the procedure in Subheading 3.1.2 with the same reagents concentration but higher number of cells seeded in 6/12/24-well plates. Cell confluency should be adjusted to your cell of interest.
14. Some suspension cells tend to get slightly attached to the plate bottom and hence pipette thoroughly while collecting cells.
15. If one's cells of interest require gentle handling, centrifugation settings can be replaced to 200 × *g* for 10 min.
16. Samples should be kept around 4 °C during preparation to prevent biological reactions from continuing.
17. It is highly important to supplement the lysis buffer with the addition of both protease and phosphatase inhibitor, as proteases and phosphatases take significant part in death signaling and might impair sample quality. Importantly, detecting pMLKL is key in this assay.
18. Generally, the sample volume that can be loaded in polyacrylamide gel electrophoresis (PAGE) is very limited. Therefore, it is very critical to adjust the volume of RIPA buffer added to the number of cells in each well so lysates from 1.5–2 × 10^5 cells can be loaded. On the other hand, loading highly concentrated lysates could disrupt the resolution of the proteins.
19. Concentrated supernatant of serum-containing medium contains massive amount of protein, frequently resulting in problematic resolution and immunoblotting. Hence the use of serum in your assay should be minimized as much as possible (with maximum of 2%).
20. Methanol and chloroform are hazardous materials and tend to evaporate. Working in a fume hood is advised. Be aware that chloroform tends to dissolve 15 and 50 mL polystyrene tubes. Therefore it is important to use 1.5 mL polypropylene tubes or glass pipettes.
21. It is important to ensure that the methanol has been completely removed so that it does not interfere with the running of the samples during PAGE.
22. Once electrophoresis is completed, do not leave your gel with samples in the tank with the power supply turned off. You might lose some of the samples by diffusion. If you want to delay your run you can decrease the voltage to 10 V and run the samples slowly to prevent diffusion of proteins.
23. We find blocking in TBS better then in PBS for the detection of phosphorylated proteins.

24. Different commercial stripping kits might have different instructions. You should take under consideration that the stripping process may damage the proteins on the membrane. Be aware that the stripping materials are often hazardous.

References

1. Kerr JF, Wyllie AH, Currie AR (1972) Apoptosis: a basic biological phenomenon with wide-ranging implications in tissue kinetics. Br J Cancer 26(4):239–257
2. Ravichandran KS, Lorenz U (2007) Engulfment of apoptotic cells: signals for a good meal. Nat Rev Immunol 7(12):964–974. https://doi.org/10.1038/nri2214
3. Fink SL, Cookson BT (2005) Apoptosis, pyroptosis, and necrosis: mechanistic description of dead and dying eukaryotic cells. Infect Immun 73(4):1907–1916. https://doi.org/10.1128/IAI.73.4.1907-1916.2005
4. Galluzzi L, Kroemer G (2008) Necroptosis: a specialized pathway of programmed necrosis. Cell 135(7):1161–1163. https://doi.org/10.1016/j.cell.2008.12.004
5. Shlomovitz ISZ, Gerlic M (2017) Mechanisms of RIPK3-induced inflammation. Immunol Cell Biol 95(2):166–172
6. Holler N, Zaru R, Micheau O, Thome M, Attinger A, Valitutti S, Bodmer JL, Schneider P, Seed B, Tschopp J (2000) Fas triggers an alternative, caspase-8-independent cell death pathway using the kinase RIP as effector molecule. Nat Immunol 1(6):489–495
7. Rickard JA, O'Donnell JA, Evans JM, Lalaoui N, Poh AR, Rogers T, Vince JE, Lawlor KE, Ninnis RL, Anderton H, Hall C, Spall SK, Phesse TJ, Abud HE, Cengia LH, Corbin J, Mifsud S, Di Rago L, Metcalf D, Ernst M, Dewson G, Roberts AW, Alexander WS, Murphy JM, Ekert PG, Masters SL, Vaux DL, Croker BA, Gerlic M, Silke J (2014) RIPK1 regulates RIPK3-MLKL-driven systemic inflammation and emergency hematopoiesis. Cell 157(5):1175–1188. https://doi.org/10.1016/j.cell.2014.04.019
8. Pasparakis M, Vandenabeele P (2015) Necroptosis and its role in inflammation. Nature 517(7534):311–320. https://doi.org/10.1038/nature14191
9. Silke J, Rickard JA, Gerlic M (2015) The diverse role of RIP kinases in necroptosis and inflammation. Nat Immunol 16(7):689–697. https://doi.org/10.1038/ni.3206
10. Taylor RC, Cullen SP, Martin SJ (2008) Apoptosis: controlled demolition at the cellular level. Nat Rev Mol Cell Biol 9(3):231–241. https://doi.org/10.1038/nrm2312
11. Oberst A, Dillon CP, Weinlich R, McCormick LL, Fitzgerald P, Pop C, Hakem R, Salvesen GS, Green DR (2011) Catalytic activity of the caspase-8-FLIP(L) complex inhibits RIPK3-dependent necrosis. Nature 471(7338):363–367. https://doi.org/10.1038/nature09852
12. Kaiser WJ, Upton JW, Long AB, Livingston-Rosanoff D, Daley-Bauer LP, Hakem R, Caspary T, Mocarski ES (2011) RIP3 mediates the embryonic lethality of caspase-8-deficient mice. Nature 471(7338):368–372. https://doi.org/10.1038/nature09857
13. Cho YS, Challa S, Moquin D, Genga R, Ray TD, Guildford M, Chan FK (2009) Phosphorylation-driven assembly of the RIP1-RIP3 complex regulates programmed necrosis and virus-induced inflammation. Cell 137(6): 1112–1123. https://doi.org/10.1016/j.cell.2009.05.037
14. He S, Wang L, Miao L, Wang T, Du F, Zhao L, Wang X (2009) Receptor interacting protein kinase-3 determines cellular necrotic response to TNF-alpha. Cell 137(6):1100–1111. https://doi.org/10.1016/j.cell.2009.05.021
15. Murphy JM, Czabotar PE, Hildebrand JM, Lucet IS, Zhang JG, Alvarez-Diaz S, Lewis R, Lalaoui N, Metcalf D, Webb AI, Young SN, Varghese LN, Tannahill GM, Hatchell EC, Majewski IJ, Okamoto T, Dobson RC, Hilton DJ, Babon JJ, Nicola NA, Strasser A, Silke J, Alexander WS (2013) The pseudokinase MLKL mediates necroptosis via a molecular switch mechanism. Immunity 39(3):443–453. https://doi.org/10.1016/j.immuni.2013.06.018
16. Sun L, Wang H, Wang Z, He S, Chen S, Liao D, Wang L, Yan J, Liu W, Lei X, Wang X (2012) Mixed lineage kinase domain-like protein mediates necrosis signaling downstream of RIP3 kinase. Cell 148(1–2):213–227. https://doi.org/10.1016/j.cell.2011.11.031
17. Hildebrand JM, Tanzer MC, Lucet IS, Young SN, Spall SK, Sharma P, Pierotti C, Garnier JM, Dobson RC, Webb AI, Tripaydonis A, Babon JJ, Mulcair MD, Scanlon MJ, Alexander WS, Wilks AF, Czabotar PE, Lessene G,

Murphy JM, Silke J (2014) Activation of the pseudokinase MLKL unleashes the four-helix bundle domain to induce membrane localization and necroptotic cell death. Proc Natl Acad Sci U S A 111(42):15072–15077. https://doi.org/10.1073/pnas.1408987111

18. Dondelinger Y, Declercq W, Montessuit S, Roelandt R, Goncalves A, Bruggeman I, Hulpiau P, Weber K, Sehon CA, Marquis RW, Bertin J, Gough PJ, Savvides S, Martinou JC, Bertrand MJ, Vandenabeele P (2014) MLKL compromises plasma membrane integrity by binding to phosphatidylinositol phosphates. Cell Rep 7(4):971–981. https://doi.org/10.1016/j.celrep.2014.04.026
19. Cai Z, Jitkaew S, Zhao J, Chiang HC, Choksi S, Liu J, Ward Y, Wu LG, Liu ZG (2014) Plasma membrane translocation of trimerized MLKL protein is required for TNF-induced necroptosis. Nat Cell Biol 16(1):55–65. https://doi.org/10.1038/ncb2883
20. Degterev A, Huang Z, Boyce M, Li Y, Jagtap P, Mizushima N, Cuny GD, Mitchison TJ, Moskowitz MA, Yuan J (2005) Chemical inhibitor of nonapoptotic cell death with therapeutic potential for ischemic brain injury. Nat Chem Biol 1(2):112–119. https://doi.org/10.1038/nchembio711
21. Welz PS, Wullaert A, Vlantis K, Kondylis V, Fernandez-Majada V, Ermolaeva M, Kirsch P, Sterner-Kock A, van Loo G, Pasparakis M (2011) FADD prevents RIP3-mediated epithelial cell necrosis and chronic intestinal inflammation. Nature 477(7364):330–334. https://doi.org/10.1038/nature10273
22. Kaiser WJ, Sridharan H, Huang C, Mandal P, Upton JW, Gough PJ, Sehon CA, Marquis RW, Bertin J, Mocarski ES (2013) Toll-like receptor 3-mediated necrosis via TRIF, RIP3, and MLKL. J Biol Chem 288(43):31268–31279. https://doi.org/10.1074/jbc.M113.462341
23. Zargarian S, Shlomovitz I, Erlich Z, Hourizadeh A, Ofir-Birin Y, Croker BA, Regev-Rudzki N, Edry-Botzer L, Gerlic M (2017) Phosphatidylserine externalization, "necroptotic bodies" release, and phagocytosis during necroptosis. PLoS Biol 15(6):e2002711. https://doi.org/10.1371/journal.pbio.2002711
24. Su Z, Yang Z, Xie L, DeWitt JP, Chen Y (2016) Cancer therapy in the necroptosis era. Cell Death Differ 23(5):748–756. https://doi.org/10.1038/cdd.2016.8
25. Mandal P, Berger SB, Pillay S, Moriwaki K, Huang C, Guo H, Lich JD, Finger J, Kasparcova V, Votta B, Ouellette M, King BW, Wisnoski D, Lakdawala AS, DeMartino MP, Casillas LN, Haile PA, Sehon CA, Marquis RW, Upton J, Daley-Bauer LP, Roback L, Ramia N, Dovey CM, Carette JE, Chan FK, Bertin J, Gough PJ, Mocarski ES, Kaiser WJ (2014) RIP3 induces apoptosis independent of pronecrotic kinase activity. Mol Cell 56(4):481–495. https://doi.org/10.1016/j.molcel.2014.10.021
26. Lawlor KE, Khan N, Mildenhall A, Gerlic M, Croker BA, D'Cruz AA, Hall C, Kaur Spall S, Anderton H, Masters SL, Rashidi M, Wicks IP, Alexander WS, Mitsuuchi Y, Benetatos CA, Condon SM, Wong WW, Silke J, Vaux DL, Vince JE (2015) RIPK3 promotes cell death and NLRP3 inflammasome activation in the absence of MLKL. Nat Commun 6:6282. https://doi.org/10.1038/ncomms7282
27. Magidson V, Khodjakov A (2013) Circumventing photodamage in live-cell microscopy. Methods Cell Biol 114:545–560. https://doi.org/10.1016/B978-0-12-407761-4.00023-3

Chapter 5

Methods for Studying TNF-Mediated Necroptosis in Cultured Cells

Zikou Liu, John Silke, and Joanne M. Hildebrand

Abstract

Necroptosis is a caspase-independent form of programmed cell death that is induced by a variety of different signalling cascades—all culminating in the activation of the pseudokinase mixed lineage kinase domain-like (MLKL). TNF-induced necroptosis is the most intensively studied of these pathways. Here we describe reagents and cell-based techniques that can be used to investigate TNF-mediated necroptosis in the lab.

Key words Necroptosis, TNF, MLKL, BN-PAGE, Cell fractionation

1 Introduction

Tumor necrosis factor (TNF) is a cytokine that plays an essential role in regulating various facets of the immune response, necroptotic cell death is one of many cellular events that can be triggered by TNF stimulation.

TNF stimulates the cell surface receptor TNF receptor I (TNFR1), which is expressed on most cells, to assemble a plasma membrane associated signaling complex. The cytosolic portion of TNFR1 recruits multiple proteins to form complex I, including cellular inhibitors of apoptosis (cIAPs), TNF receptor associated death domain (TRADD), receptor interacting serine/threonine protein kinase 1 (RIPK1) and TNFR-associated factor 2 (TRAF2) [1, 2]. The E3 ligase, cIAP1, catalyzes the Lys63-linked polyubiquitylation of several proteins including RIPK1 to form a docking site for the Linear UBiquitin chain Assembly Complex (LUBAC) and subsequent activation of the Nuclear Factor-κB (NF-κB) and MAPK signaling pathways [3]. Activation of these pathways results in production of $cFLIP_L$, a caspase-8 inhibitor, which helps prevent TNF induced apoptosis.

When cIAPs are depleted or suppressed by genetic or pharmacological means, e.g., by using SMAC-mimetics like birinapant [4], the activation of NF-κB and MAPK pathways are perturbed

Adrian T. Ting (ed.), *Programmed Necrosis: Methods and Protocols*, Methods in Molecular Biology, vol. 1857, https://doi.org/10.1007/978-1-4939-8754-2_5,

and nonubiquitylated RIPK1 favors the formation of complex II which contains TRADD, Fas-associated protein with death domain (FADD) and caspase-8 [1]. In complex II, the lack of cFLIP$_L$, allows caspase-8 to become fully activated and activate downstream effector caspases, such as caspase-3 to cause an apoptotic cell death [5, 6]. Caspase-8 also cleaves RIPK1 preventing it from activating the necroptotic cell death pathway.

If however caspase-8 is inhibited by caspase inhibitors, such as Q-VD-OPh [7], Z-VAD-FMK or emricasan (IDN-6556) [4], the RIP homotypic interaction motif (RHIM) in RIPK1 binds to the RHIM in RIPK3 [8], facilitating RIPK3 oligomerization and phosphorylation [9]. Phosphorylated RIPK3 then phosphorylates mixed lineage kinase domain-like protein (MLKL) within the necrosome, also known as complex II(n) [4, 10].

In the necrosome, RIPK3 phosphorylates T357/S358 of human MLKL and S345 of mouse MLKL, triggering MLKL oligomerization and its association with biological membranes [11, 12]. While the precise mode of membrane disruption is still a matter of contention, it is generally believed that the four helix bundle domain (4HB) of MLKL is absolutely required [13–15].

Each of these signaling events can be induced in a wide variety of laboratory-cultured cells using combinations of commercially available recombinant stimulatory cytokines, small-molecule inhibitors of cIAPs (or "SMAC mimetics") and caspase inhibitors. Here we provide instructions and recommendations for the use of such reagents, and the measurement of three main events in this signaling pathway: the phosphorylation of MLKL, the assembly of high molecular weight MLKL-containing complexes [16, 17] and the disruption of cellular membranes.

2 Materials

2.1 Inducing Necroptosis in Cultured Cell Lines

1. Cell lines: immortalized wild type Mouse Dermal Fibroblasts (MDF) cultured in Dulbecco's Modified Eagle Medium (DMEM) + 8% Fetal Calf Serum (FCS) at 37 °C, 10% CO_2.
2. Sterile Dulbecco's phosphate-buffered saline (DPBS) and trypsin–EDTA. These are used to detach the adherent MDF cell line for counting and maintenance.
3. TNF: pure recombinant TNF (100 ng/mL) used to induce necroptosis in the presence of SMAC mimetics and caspase inhibitors. We use hTNF-Fc made in-house for experiments in both mouse and human derived cell lines. Recombinant TNF is available from commercial suppliers.
4. SMAC mimetics: 500 nM birinapant or 500 nM Compound A [18] (*see* **Note 1**). Birinipant and other SMAC mimetic compounds are available from commercial suppliers.

5. Caspase inhibitors: 5 μM Q-VD-OPh, 20 μM Z-VAD-FMK, or 5 μM emricasan (IDN-6556) (*see* **Note 1**). All of these reagents are available from commercial suppliers.
6. 2 × SDS-PAGE loading buffer: 126 mM Tris–HCl pH 8, 20% v/v glycerol, 4% w/v SDS, 0.02% w/v bromophenol blue, 5% v/v 2-mercaptoethanol.
7. Home-made or commercial precast SDS-PAGE gel. Nitrocellulose membrane and buffers for gel electrophoresis, Western transfer, Western blotting and chemiluminescent detection.
8. Propidium iodide: 1.0 mg/mL for 1000× solution (*see* **Note 2**).
9. Mouse phospho-MLKL (Ser345) rabbit monoclonal EPR9515(2) (Abcam).
10. MLKL rat monoclonal 3H1 (Millipore).
11. Cleaved CASPASE 8 (Asp387) rabbit monoclonal D5B2 (Cell Signaling Technology).
12. β-ACTIN mouse monoclonal AC-15 (Sigma-Aldrich).
13. BAK NT rabbit polyclonal #06-536 (EMD Millipore).
14. GAPDH rabbit monoclonal 14C10 (Cell Signaling Technology).
15. HRP-conjugated secondary antibodies.

2.2 Detecting the Formation of High Molecular Weight MLKL-Containing Complexes in Cellular Membranes

1. MELB buffer: 20 mM HEPES pH 7.5, 100 mM KCl, 2.5 mM $MgCl_2$, and 100 mM sucrose (*see* **Note 3**). Use for cell lysis and membrane solubilisation.
2. Digitonin: Make 10% w/v stock in Milli-Q water.
3. DUB inhibitor: N-ethyl maleimide (NEM), store at 1 M at −20 °C, use at 2 mM.
4. Protease inhibitor cocktail: e.g., Complete Protease Inhibitor tablets (Roche) used as recommended by manufacturer.
5. Phosphatase inhibitors: 5 mM ß-glycerophosphate, 1 mM sodium molybdate, 2 mM sodium pyrophosphate (*see* **Note 4**).
6. BN Lysis buffer: MELB buffer, 0.025% digitonin, protease inhibitor cocktail, phosphatase inhibitors and NEM.
7. BN Membrane solubilisation buffer: MELB buffer, 1% digitonin, protease inhibitor cocktail, phosphatase inhibitors and NEM.
8. 10× Blue-native PAGE loading buffer:5% Coomassie Blue G, 500 mM ε-amino n-caproic acid, 100 mM Bis–Tris pH 7.0.
9. Homemade or commercial precast Bis–Tris 4–16% Native-PAGE gel and running buffers.
10. PVDF membrane.
11. BN-PVDF denaturing solution: 6 M guanidine–HCl, 10 mM Tris–HCl pH 7.0.
12. BN-PVDF destain: 50% methanol and 25% acetic acid in Milli-Q water.

3 Methods

3.1 Measuring Necroptosis Using Flow Cytometry

1. Seed MDFs in 24-well plates at 50,000 cells/well, using 500 μL DMEM + 8% FCS. Allow for 2 wells per condition, one for flow cytometry quantification of cell death and one for Western blot analysis. Allow at least 3 h for cells to reattach before stimulating.
2. Prior to stimulation, remove media and any unattached cells and replace with 500 μL of fresh prewarmed media. Add TNF, SMAC-mimetic, and caspase inhibitor as required, being sure to control for DMSO diluent in "untreated" control wells. Incubate at 37 °C, 10% CO_2.
3. Harvest all cells after 3 or 25 h from each well by retaining the media (which will contain dead and floating cells), washing the remaining attached cells with DPBS and then using trypsin–EDTA to detach live adherent cells from the plate. The media and the trypsin–EDTA components are combined in the same collecting tube.
4. After centrifuging at 1000 × g for 5 min, the supernatant is carefully discarded and the cell pellet is resuspended in 100 μL of DPBS containing 1 μg/mL propidium iodide (*see* **Note 5**) for flow cytometric analysis (Fig. 1) or 100 μL SDS-PAGE sample loading buffer for Western blot analysis (Fig. 2).

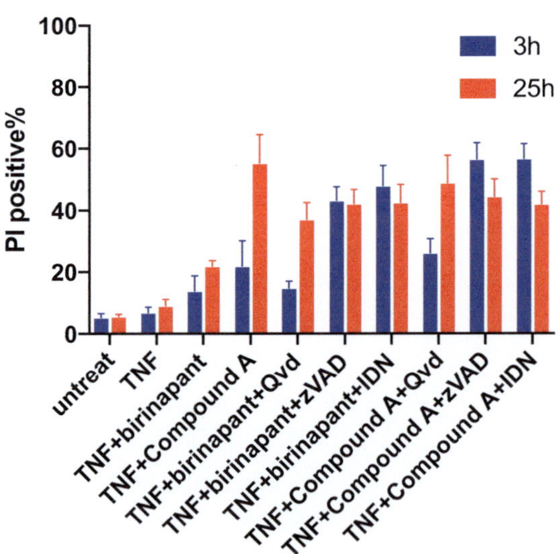

Fig. 1 Wild type mouse dermal fibroblasts (MDF) were treated for 3 or 25 h using TNF and varying combinations of two commonly used SMAC mimetic compounds and three commonly used pan-caspase inhibitors for apoptotic and necroptotic stimulation respectively. Cell death was measured using PI staining and flow cytometry. Data are plotted as mean +/− SEM of three independent experiments (*see* **Note 8**)

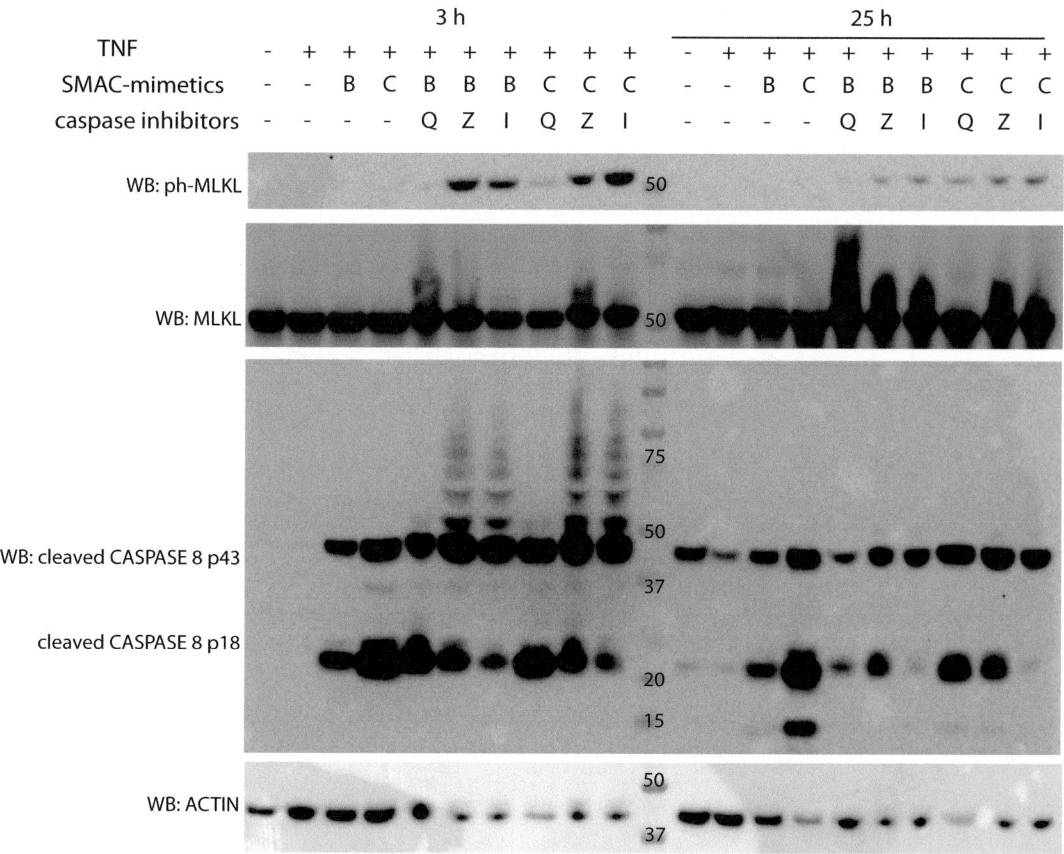

Fig. 2 Whole cell lysates were separated by reducing SDS-PAGE to demonstrate western-blot detection of phosphorylated MLKL, total MLKL, cleaved CASPASE 8 and actin (*see* **Note 9**). (*B* birinapant, *C* Compound A, *Q* Q-VD-OPH, *Z* zVAD-FMK, *I* IDN-6556 (emricasan))

3.2 Detecting the Formation of High Molecular Weight MLKL-Containing Complexes in Cellular Membranes

3.2.1 Fractionation of Cells into Cytoplasmic and Crude–Membrane Fractions

Carry out all procedures on ice, unless otherwise specified.

1. Plates 500,000 MDFs per well in 6-well plates, (allowing one well per condition) and wait at least 3 h until they are attached.
2. Treat cells with combined necroptotic stimuli T + C + I at indicated dosage for 0, 1, 2, and 3 h.
3. Use a cell scraper to detach living adherent cells from the plate, and remove entire contents of each well (media containing detached and dead cells, mechanically detached living cells) to a 10 mL Falcon tubes, spin at 1000 × *g* for 5 min. Remove the supernatant and resuspend cell pellet in 1 mL cold DPBS, transfer to 1.5 mL microfuge tubes. Spin again at 3500 × *g* for 2 min, and discard wash.
4. Resuspend cell pellet in 200 μL BN lysis buffer (*see* **Note 6**), pipetting up and down 10 times, and leave on ice for 10–20 min. Set aside 20 μL as the "whole cell lysate" sample for SDS PAGE.

5. Spin the lysate at 15,000 × *g* for 10 min at 4 °C. Transfer supernatant to a fresh tube—this is your "cytoplasm" fraction. Combine 45 μL cytoplasmic fraction with 5 μL 10% digitonin to give final concentration of 1% v/v for BN PAGE.
6. Add 200 μL MELB buffer to each tube, being careful not to disturb the pellet to gently rinse off and remove remnant cytoplasm fraction on tube sides.
7. Add 100 μL BN membrane solubilisation buffer to each pellet and pipet at least 15 times until fully resuspended. Leave on ice for 20 min with intermittent vortexing.
8. Centrifuge 20,000 × *g* for 10 min at 4 °C. The supernatant contains the 1% digitonin soluble membrane fraction for BN-PAGE.

3.2.2 Blue Native-PAGE, Western Transfer and Western Blotting

1. Mix 22.5 μL of cytoplasm or membrane fraction samples with 2.5 μL BN-PAGE loading buffer (10×) for each time point. Load and run BN gel according to the manufacturers' instructions.
2. Set up Western transfer onto PVDF (nitrocellulose membrane is less stable in BN-destain solution) adding 0.0375% SDS to the transfer buffer to assist with further protein denaturation. Transfer at 50 V for 3 h.
3. Destain PVDF membrane for 2 min being sure to pencil in BN molecular weight markers as soon as they appear. Continue destaining until there is no change in the level of blue.
4. Incubate PVDF membrane in denaturing solution for 30 min at room temp in a sealed container, this can help to further unmask protein epitopes for Western detection by some antibodies (*see* **Note 7**)
5. Rinse the membrane with water and then with PBS-T. Then block with 2% w/v skim milk in PBS-T for more than 1 h. Proceed to western blot with anti-phMLKL and anti-MLKL sequentially (Fig. 3).

4 Notes

1. B: birinapant, C: Compound A, Q: Q-VD-OPH, Z: zVAD-FMK, and I: IDN-6556.
2. PI; propidium iodide in the following text. PI stock should be wrapped in foil to prevent light-induced decomposition and diluted in DPBS at 1:1000 prior to use.
3. MELB buffer needs to be frozen between uses to prevent microbial growth.
4. Phosphatase inhibitors are dissolved in Milli-Q water as 100× stock and frozen for storage.
5. PI fluorescence can be quantified using the FL3 channel on any standard flow cytometer. Use an appropriate FSC/SSC gating to exclude cell debris. Typically cell debris has a much

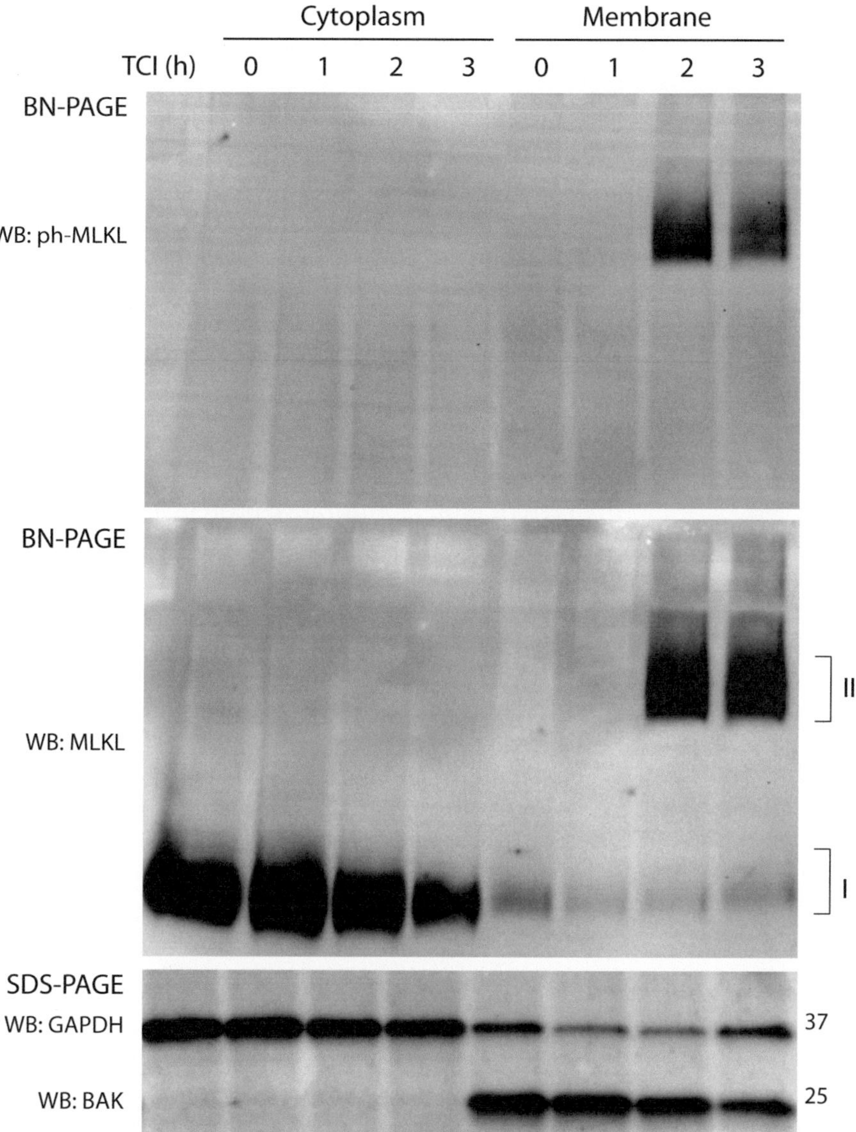

Fig. 3 Wild type MDF were treated with TCI for 1, 2 or 3 h and crude cytoplasmic and membrane fractions were prepared. These fractions were separated by nonreducing Blue-Native PAGE. Phosphorylated MLKL and total MLKL were detected by western blot. Western detection of GAPDH (cytosolic) and BAK (mitochondrial membrane) were used to demonstrate the quality of cell fractionation (*see* **Note 10**)

smaller Forward Scatter (FSC) and Side Scatter (SSC) than the corresponding living cell and because a single dead cell could generate several debris particles it is not appropriate to include the debris in an analysis as it would tend to bias results in favor of "dead cells." Events plotted as FL3 vs. FSC should fall into two easily defined populations, FL3 High (PI permeant, dead) and FL3 low (PI impermeant, live). Use appropriate software to quantify PI permeant cells as a proportion of all intact cells.

6. We recommend the addition of DUB, phosphatase, and protease inhibitors to BN lysis buffer fresh each time the experiment is performed to ensure they are maximally effective.
7. The MLKL antibody is generated against amino acid residues 126–180 from murine MLKL which encompass the brace region of MLKL. BN-PAGE maintains proteins and protein complexes in their native state and we have found that the use of denaturing solution facilitates improved MLKL detection.
8. At the doses used here, in MDFs, we find Z-VAD-FMK and emricasan induce more robust necroptosis than Q-VD-OPh after 3 h, but no significant difference between caspase inhibitors after 22 h.
9. It is important to take cell death and the loss of cytoplasmic contents into account when measuring relative protein levels by Western blot. Comparative protein levels should be considered relative to the levels of actin detectable. After 3 h of treatment, the amounts of phosphorylated MLKL detected by Western blot correlate positively with the levels of cell death induced. Interestingly, after 22 h, the levels of anti-phMLKL detectable by Western blot diminish significantly while total MLKL levels do not, suggesting that phMLKL is selectively less stable in cells. As published previously [4], emricasan is the most potent inhibitor of CASPASE 8- CFLIPL heterodimers cleavage when compared to Q-VD-OPh and Z-VAD-FMK at the doses tested. Despite significant CASPASE 8 activity being detected after 3 h treatments with TNF, Smac-mimetic, and Z-VAD-FMK—significant MLKL phosphorylation was still detectable.
10. A high molecular weight, membrane associated MLKL-containing complex is detectable as early as 2 h following stimulation of MDF with TNF, Compound A, and emricasan. Detection of the cytoplasmic marker GAPDH suggests that there is some contamination of the crude membrane fraction with cytosolic contents, explaining the detection of a lower molecular weight MLKL species in membrane fractions.

Acknowledgment

We wish to acknowledge the Australian National Health and Medical Research Council (GNT1105023, GNT1107149 and GNT1142669) and the Melbourne International Research Scholarship for funding this work.

References

1. Micheau O, Tschopp J (2003) Induction of TNF receptor I-mediated apoptosis via two sequential signaling complexes. Cell 114(2): 181–190
2. Silke J, Brink R (2010) Regulation of TNFRSF and innate immune signalling complexes by TRAFs and cIAPs. Cell Death Differ 17(1):35–45. https://doi.org/10.1038/cdd.2009.114
3. Bertrand MJ, Milutinovic S, Dickson KM, Ho WC, Boudreault A, Durkin J, Gillard JW, Jaquith JB, Morris SJ, Barker PA (2008) cIAP1 and cIAP2 facilitate cancer cell survival by functioning as E3 ligases that promote RIP1 ubiquitination. Mol Cell 30(6):689–700. https://doi.org/10.1016/j.molcel.2008.05.014
4. Brumatti G, Ma C, Lalaoui N, Nguyen NY, Navarro M, Tanzer MC, Richmond J, Ghisi M, Salmon JM, Silke N, Pomilio G, Glaser SP, de Valle E, Gugasyan R, Gurthridge MA, Condon SM, Johnstone RW, Lock R, Salvesen G, Wei A, Vaux DL, Ekert PG, Silke J (2016) The caspase-8 inhibitor emricasan combines with the SMAC mimetic birinapant to induce necroptosis and treat acute myeloid leukemia. Sci Transl Med 8(339):339–ra369. https://doi.org/10.1126/scitranslmed.aad3099
5. Lin Y, Devin A, Rodriguez Y, Liu ZG (1999) Cleavage of the death domain kinase RIP by caspase-8 prompts TNF-induced apoptosis. Genes Dev 13(19):2514–2526
6. He S, Wang L, Miao L, Wang T, Du F, Zhao L, Wang X (2009) Receptor interacting protein kinase-3 determines cellular necrotic response to TNF-alpha. Cell 137(6):1100–1111. https://doi.org/10.1016/j.cell.2009.05.021
7. Caserta TM, Smith AN, Gultice AD, Reedy MA, Brown TL (2003) Q-VD-OPh, a broad spectrum caspase inhibitor with potent antiapoptotic properties. Apoptosis 8(4):345–352
8. Sun X, Yin J, Starovasnik MA, Fairbrother WJ, Dixit VM (2002) Identification of a novel homotypic interaction motif required for the phosphorylation of receptor-interacting protein (RIP) by RIP3. J Biol Chem 277(11):9505–9511. https://doi.org/10.1074/jbc.M109488200
9. Wu XN, Yang ZH, Wang XK, Zhang Y, Wan H, Song Y, Chen X, Shao J, Han J (2014) Distinct roles of RIP1-RIP3 hetero- and RIP3-RIP3 homo-interaction in mediating necroptosis. Cell Death Differ 21(11):1709–1720. https://doi.org/10.1038/cdd.2014.77
10. Murphy JM, Silke J (2014) Ars Moriendi; the art of dying well—new insights into the molecular pathways of necroptotic cell death. EMBO Rep 15(2):155–164. https://doi.org/10.1002/embr.201337970
11. Sun L, Wang H, Wang Z, He S, Chen S, Liao D, Wang L, Yan J, Liu W, Lei X, Wang X (2012) Mixed lineage kinase domain-like protein mediates necrosis signaling downstream of RIP3 kinase. Cell 148(1-2):213–227. https://doi.org/10.1016/j.cell.2011.11.031
12. Murphy JM, Czabotar PE, Hildebrand JM, Lucet IS, Zhang JG, Alvarez-Diaz S, Lewis R, Lalaoui N, Metcalf D, Webb AI, Young SN, Varghese LN, Tannahill GM, Hatchell EC, Majewski IJ, Okamoto T, Dobson RC, Hilton DJ, Babon JJ, Nicola NA, Strasser A, Silke J, Alexander WS (2013) The pseudokinase MLKL mediates necroptosis via a molecular switch mechanism. Immunity 39(3):443–453. https://doi.org/10.1016/j.immuni.2013.06.018
13. Cai Z, Jitkaew S, Zhao J, Chiang HC, Choksi S, Liu J, Ward Y, Wu LG, Liu ZG (2014) Plasma membrane translocation of trimerized MLKL protein is required for TNF-induced necroptosis. Nat Cell Biol 16(1):55–65. https://doi.org/10.1038/ncb2883
14. Chen X, Li W, Ren J, Huang D, He WT, Song Y, Yang C, Li W, Zheng X, Chen P, Han J (2014) Translocation of mixed lineage kinase domain-like protein to plasma membrane leads to necrotic cell death. Cell Res 24(1):105–121. https://doi.org/10.1038/cr.2013.171
15. Xia B, Fang S, Chen X, Hu H, Chen P, Wang H, Gao Z (2016) MLKL forms cation channels. Cell Res 26(5):517–528. https://doi.org/10.1038/cr.2016.26
16. Hildebrand JM, Tanzer MC, Lucet IS, Young SN, Spall SK, Sharma P, Pierotti C, Garnier JM, Dobson RC, Webb AI, Tripaydonis A, Babon JJ, Mulcair MD, Scanlon MJ, Alexander WS, Wilks AF, Czabotar PE, Lessene G, Murphy JM, Silke J (2014) Activation of the pseudokinase MLKL unleashes the four-helix bundle domain to induce membrane localization and necroptotic cell death. Proc Natl Acad Sci U S A 111(42):15072–15077. https://doi.org/10.1073/pnas.1408987111
17. Schagger H, von Jagow G (1991) Blue native electrophoresis for isolation of membrane protein complexes in enzymatically active form. Anal Biochem 199(2):223–231
18. Vince JE, Wong WW, Khan N, Feltham R, Chau D, Ahmed AU, Benetatos CA, Chunduru SK, Condon SM, McKinlay M, Brink R, Leverkus M, Tergaonkar V, Schneider P, Callus BA, Koentgen F, Vaux DL, Silke J (2007) IAP antagonists target cIAP1 to induce TNFalpha-dependent apoptosis. Cell 131(4):682–693. https://doi.org/10.1016/j.cell.2007.10.037

Chapter 6

Analysis of Necroptosis in Bone Marrow-Derived Macrophages

Diana Legarda and Adrian T. Ting

Abstract

Necroptosis in macrophages is increasingly being recognized as a significant contributor to the inflammation seen in many pathologies. Here we describe a well-known method to obtain quiescent, mature macrophages that can be used to study the molecular mechanisms that regulate necroptosis. We also describe two ways to quantify this form of programmed cell death.

Key words Bone marrow-derived macrophages, Necroptosis, Cell death, Propidium iodide exclusion, ATP viability assay, Luminescence

1 Introduction

Necroptosis is a highly inflammatory mode of programmed cell death that has been implicated in the pathology of many diseases such as atherosclerosis, inflammatory bowel disease, viral and bacterial infection, neurodegeneration, and some cancers [1]. The molecular mechanisms that regulate this form of cell death has been well characterized downstream of the tumor necrosis factor (TNF) receptor [2, 3]. Since macrophages play a critical role in the progression of inflammation, host defense, and immunity, it is important to understand the regulation of necroptosis in these innate immune cells. We and others have previously demonstrated that macrophages undergo necroptosis induce by lipopolysaccharide (LPS), a component of the bacterial cell wall, and this is accomplished through induction of TNF and interferon, and cross talk between the signaling pathways downstream of their respective receptors [4–7].

In this chapter, we describe the isolation of macrophage progenitor cells from mouse femur and tibia, and their differentiation into mature macrophages [8]. Caspase-8 activity can suppress RIPK1 and RIPK3-dependent necroptosis [9, 10], and so

Adrian T. Ting (ed.), *Programmed Necrosis: Methods and Protocols*, Methods in Molecular Biology, vol. 1857, https://doi.org/10.1007/978-1-4939-8754-2_6, © Springer Science+Business Media, LLC, part of Springer Nature 2018

suppression of caspase activity is required to unleash necroptosis. In our studies, we pretreat macrophages with the cell-permeable pancaspase inhibitor z-VAD-FMK prior to LPS treatment to induce necroptosis. A hallmark morphological feature of necroptotic cells is loss of membrane integrity. To quantify this form of cell death, we utilize this feature by using propidium iodide (PI), a fluorescent dye that is excluded from viable cells but is taken up by dying cells whose membrane integrity is compromised. A reduction in ATP levels can be an indication of a reduction of viable, intact cells because of necroptosis. Therefore, we also use a commercially available cell viability assay that measures ATP levels to quantify necroptosis.

2 Materials

2.1 Mouse Dissection

1. Surgical scissors.
2. 70% ethanol.
3. 1× PBS.

2.2 Bone Marrow-Derived Macrophage (BMDM) Progenitor Isolation from Femur and Tibia of a Mouse

1. 27-gauge needle.
2. 3 mL syringe.
3. Razor.
4. Sterile surgical scissors.
5. 50 mL conical tube.
6. 0.45 μm filter.
7. Tissue culture and non-tissue culture treated 10 cm plates.

2.3 Generation of L929 Conditioned Media

1. L929 (ATCC).
2. T75 tissue culture flasks.
3. 0.45 μm filter.

2.4 Cell Culture

1. Humidified incubator at 37 °C with 5% CO_2.
2. Hemocytometer.
3. RPMI complete: RPMI 1640, 10% fetal calf serum (FCS), 100 IU penicillin 100 μg/mL streptomycin solution, 15 μg/mL gentamicin, 1% nonessential amino acids, and 0.1% beta-mercaptoethanol.
4. 0.25% trypsin containing 2.21 mM EDTA.
5. 24-well non-tissue culture treated plates.
6. 96-well tissue culture treated plates.
7. 15 and 50 mL conical tubes.
8. Z-Val-ala-asp-(OMe)-fluoromethyl ketone (z-VAD-FMK).
9. Lipopolysaccharide (LPS) from *E. coli* 0111:B4.

2.5 Analysis of Necroptosis by Propidium Iodide Exclusion

1. 1 mg/mL propidium iodide.
2. 1× PBS.
3. 5 mL polypropylene round-bottom tubes.
4. Flow cytometer, such as BD LSR Fortessa™ or BD Accuri™ C6 (BD biosciences).
5. Flow cytometry analysis software, such as FlowJo®.

2.6 Analysis of Necroptosis by ATP Levels

1. CellTiter-Glo® Luminescent Cell Viability Assay (Promega).
2. Solid, white 96-well plates.
3. Microplate reader, such as POLARstar® omega (BMG Labtech).

3 Methods

3.1 Preparation of L929-Conditioned Medium (L929-CM)

1. Plate 5×10^5 of L929 cells in a T75 cm flask containing 50 mL of RPMI complete media.
2. Grow at 37°C in a humidified incubator with 5% CO_2 for 7 days.
3. Remove conditioned medium from L929 cells and filter through a 0.45 μm filter.
4. Store the L929-conditioned medium in 50 mL aliquots at −80°C.

3.2 Preparation and Differentiation of BMDMs

1. Euthanize the mouse by CO_2 asphyxiation followed by cervical dislocation.
2. Sterilize abdomen and hind legs with 70% ethanol.
3. Make an incision in the abdomen and cut outward to expose the hind legs.
4. Cut the femur at the proximal end while keeping the epiphysis intact, and the tibia at the distal end. Do not expose the contents. Cut away the surrounding muscle.
5. Dip the femur and tibia in 70% ethanol and allow them to air-dry.
6. Using sterile techniques, cut the femur at the proximal epiphysis and at the knee joint. Using a 27-gauge needle and 3 mL syringe, flush the femur with 1× PBS. Repeat with the tibia. The lumen of the femur and tibia should go from red to white as the bone marrow cells are flushed out.
7. Break up the bone marrow clumps and bring to a single cell suspension using the needle and syringe.
8. Pass the cells through a cell strainer. Wash with an additional 5 mL of 1× PBS by centrifuging at $250 \times g$ for 5 min and removing the supernatant.

9. Resuspend the cells in 7 mL of RPMI complete media and 3 mL of L929 conditioned media (L929-CM) so that the final concentration of L929-CM is 30%.
10. Count the cells with a hemocytometer and plate the cells at a density of 10×10^6 per 10 cm tissue culture treated plate, using RPMI complete containing 30% L929-CM (*see* **Note 1**). Place in a humidified incubator at 37 °C with 5% CO_2, overnight.
11. The next day, replate the supernatant from one plate into three 10 cm non-tissue culture treated plates. Bring the volume in each plate to 10 mL using 7 mL of RPMI complete and 3 mL of L929-CM, so that the final concentration of L929-CM is 30%. The supernatant will consist of bone marrow progenitor cells depleted of other adherent cells, such as fibroblasts.
12. On day 4, supplement the media with 3.5 mL of RPMI complete plus 1.5 mL of L929-CM, so that the final concentration of L929-CM is 30%.
13. Allow the BMDMs to differentiate for 7–10 days in a humidified incubator at 37 °C with 5% CO_2.

3.3 Propidium Iodide Exclusion Assay

1. Remove the cells from 10 cm non-tissue culture treated plates using 1 mL 0.25% trypsin containing 2.21 mM EDTA per plate.
2. Incubate at 37 °C with 5% CO_2 for approximately 5 min, or until the cells are lifted from the plate (*see* **Note 2**).
3. Deactivate the trypsin by adding 10 mL of RPMI complete media to the plate and gently pipetting to dislodge cells from the plate.
4. Transfer to a conical tube and centrifuging at $250 \times g$ for 5 min.
5. Wash the cells by resuspending them in 10 mL of 1× PBS, centrifuging at $250 \times g$ for 5 min, and removing the supernatant.
6. Plate the BMDMs in a 24-well non-tissue culture treated plate at a density of 0.25×10^6 per well in 0.5 mL of RPMI complete containing 30% L929-CM. Each experimental condition should be carried out in triplicate. Allow the cells to attach to the bottom of the well by culturing for an additional 24 h.
7. Pretreat the cells with 25 μm of z-VAD-FMK for 30 min. Stimulate the cells with LPS from *E. coli* 0111:B4, at doses ranging from 0.1 ng/mL to 1 ng/mL for 24 h.
8. After stimulation, transfer the media from the 24-well plate to 5 mL polypropylene round-bottom tubes (*see* **Note 3**).
9. Wash the adherent cells in 1 mL of 1× PBS. Remove the cells from the plate by adding 200 μL of 0.25% trypsin–EDTA.
10. Incubate at 37 °C with 5% CO_2 for approximately 5 min, or until the cells are lifted from the plate.

11. Add 500 μL of 1× PBS and transfer the entire contents of the well to the corresponding 5 mL polypropylene tube in **step 5** (*see* **Note 3**). Bring the volume in each tube to approximately 4 mL with 1× PBS.
12. Centrifuge at 250 × *g* for 5 min and carefully remove the supernatant.
13. To the cell pellet, add 100 μL of propidium iodide that has been diluted 1:100 in 1× PBS. Incubate at room temperature for 5–10 min. Then add 400 μL of 1× PBS to each tube.
14. Analyze propidium iodide uptake by flow cytometry (*see* **Note 4**). Results from a typical experiment are shown in Fig. 1.

3.4 Viability Assay

1. Remove the cells from 10 cm non-tissue culture treated plates using 1 mL 0.25% trypsin–EDTA.
2. Incubate at 37 °C with 5% CO_2 for approximately 5 min, or until the cells are lifted from the plate.
3. Wash the cells in 10 mL of 1× PBS.
4. Plate the cells in 96-well tissue culture treated plates at a density of 25,000 cells per well in 50 μL of RPMI complete containing 30% L929-CM. Each experimental condition should be carried out in triplicate. Allow the cells to incubate for an additional 24 h.
5. Pretreat the cells with 25 μm of z-VAD-FMK for 30 min. Stimulate the cells with LPS from *E. coli* 0111:B4, at doses ranging from 0.1 ng/mL to 1 ng/mL for 24 h. Note that the final volume in each well of the 96-well plate should be 100 μL.

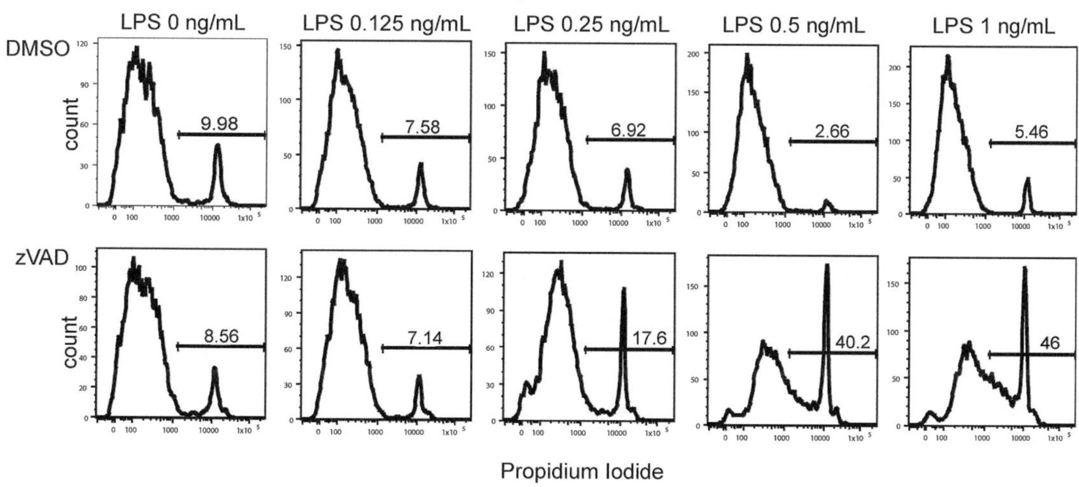

Fig. 1 BMDMs were pretreated with 25 μm z-VAD-FMK or DMSO prior to stimulation with the indicated concentrations of LPS for 24 h. Cell death was measured by propidium iodide exclusion and flow cytometry analysis. The experiment shown is representative of at least three independent experiments

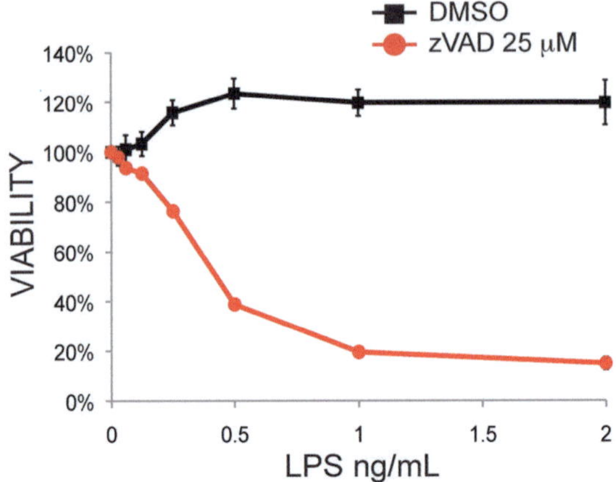

Fig. 2 BMDMs were pretreated with 25 μm z-VAD-FMK or DMSO prior to stimulation with the indicated concentrations of LPS for 24 h. Cell viability was determined using CellTiter-Glo® Luminescent Cell Viability Assay. Data is presented as the mean of an experiment performed in triplicate, ± standard deviation. The experiment show is representative of at least three independent experiments

6. Prepare reagents in the CellTiter-Glo® Luminescent Cell Viability Assay according to the manufacturer's instruction (*see* **Note 5**).
7. After stimulation, allow reagent and 96-well plate to equilibrate to room temperature (approximately 30 min). Then add an equal volume of the reagent to each well (*see* **Note 6**).
8. Mix on an orbital shaker for 2 min. Incubate the plate at room temperature for 10 min.
9. Transfer the contents of the 96-well plate to a solid, white 96-well plate.
10. Record the luminescence on a plate reader.
11. Cell viability is calculated as a percentage relative to unstimulated control cells (100% viability). A representative experiment is shown in Fig. 2.

4 Notes

1. Alternatively, bone marrow progenitor cells can be frozen in 90% FCS and 10% dimethylsulfoxide (DMSO) at a density of 4 to 6 × 10^6 cells/mL. Transfer 1 mL of the freezing media/cell suspension to each cryovial and place on ice for 10 min. Store in −80°C for 24–48 h and transfer to liquid nitrogen container for long-term storage. To thaw, place the frozen vial in a 37 °C water bath for about 10 min. Wash the cells with 1× PBS, resuspend in complete media, and plate as usual [11].

2. BMDMs are difficult to trypsinize. Therefore, it may take longer than 5 min of incubation with trypsin–EDTA to lift the BMDMs from the plate.
3. Cells that have died off can float off the plate and these are collected first. Then the adherent cells still stuck to the plate are removed by trypsin and pooled together with those that have floated off. This is to ensure that those cells that die earlier are not lost in the analysis.
4. Propidium iodide is excited by wavelengths between 400 and 600 nm and emits light between 600 and 700 nm.
5. Aliquots of CellTiter-Glo® Luminescent Cell Viability Assay can be stored at −20°C until use.
6. We have found that using the reagent diluted in a ratio of 1:4 with the cell culture media is sufficient to lyse the cells. In this scenario, the cells can be plated in a total volume of 200 μL per well, rather than 100 μL.

Acknowledgments

This work was supported by grants AI052417 and DK072201 from the NIH, and a Senior Research Award from the Crohn's and Colitis Foundation of America (CCFA).

References

1. Zhou W, Yuan J (2014) Necroptosis in health and diseases. Semin Cell Dev Biol 35:14–23. https://doi.org/10.1016/j.semcdb.2014.07.013
2. Justus SJ, Ting AT (2015) Cloaked in ubiquitin, a killer hides in plain sight: the molecular regulation of RIPK1. Immunol Rev 266(1):145–160. https://doi.org/10.1111/imr.12304
3. O'Donnell MA, Ting AT (2012) NFkappaB and ubiquitination: partners in disarming RIPK1-mediated cell death. Immunol Res 54(1-3):214–226. https://doi.org/10.1007/s12026-012-8321-7
4. He S, Liang Y, Shao F, Wang X (2011) Toll-like receptors activate programmed necrosis in macrophages through a receptor-interacting kinase-3-mediated pathway. Proc Natl Acad Sci U S A 108(50):20054–20059. https://doi.org/10.1073/pnas.1116302108
5. Legarda D, Justus SJ, Ang RL, Rikhi N, Li W, Moran TM, Zhang J, Mizoguchi E, Zelic M, Kelliher MA, Blander JM, Ting AT (2016) CYLD proteolysis protects macrophages from TNF-mediated auto-necroptosis induced by LPS and licensed by type I IFN. Cell Rep 15(11):2449–2461. https://doi.org/10.1016/j.celrep.2016.05.032
6. McComb S, Cessford E, Alturki NA, Joseph J, Shutinoski B, Startek JB, Gamero AM, Mossman KL, Sad S (2014) Type-I interferon signaling through ISGF3 complex is required for sustained Rip3 activation and necroptosis in macrophages. Proc Natl Acad Sci U S A 111(31):E3206–E3213. https://doi.org/10.1073/pnas.1407068111
7. Robinson N, McComb S, Mulligan R, Dudani R, Krishnan L, Sad S (2012) Type I interferon induces necroptosis in macrophages during infection with salmonella enterica serovar typhimurium. Nat Immunol 13(10):954–962. https://doi.org/10.1038/ni.2397
8. Zhang X, Goncalves R, Mosser DM (2008) The isolation and characterization of murine

macrophages. Curr Protoc Immunol Chapter 14:Unit 14 11. doi:https://doi.org/10.1002/0471142735.im1401s83

9. O'Donnell MA, Perez-Jimenez E, Oberst A, Ng A, Massoumi R, Xavier R, Green DR, Ting AT (2011) Caspase 8 inhibits programmed necrosis by processing CYLD. Nat Cell Biol 13(12):1437–1442. https://doi.org/10.1038/ncb2362

10. Oberst A, Dillon CP, Weinlich R, McCormick LL, Fitzgerald P, Pop C, Hakem R, Salvesen GS, Green DR (2011) Catalytic activity of the caspase-8-FLIP(L) complex inhibits RIPK3-dependent necrosis. Nature 471(7338):363–367. https://doi.org/10.1038/nature09852

11. Marim FM, Silveira TN, Lima DS Jr, Zamboni DS (2010) A method for generation of bone marrow-derived macrophages from cryopreserved mouse bone marrow cells. PLoS One 5(12):e15263. https://doi.org/10.1371/journal.pone.0015263

Chapter 7

Generation and Use of Chimeric RIP Kinase Molecules to Study Necroptosis

Diego A. Rodriguez and Douglas R. Green

Abstract

Necroptosis, a form of regulated necrosis, is triggered by a variety of signals that converge to activate receptor interacting protein kinase-3 (RIPK3), consequently promoting the direct phosphorylation and activation of the mixed lineage kinase like (MLKL) protein. Active MLKL executes necroptosis by disrupting the integrity of the plasma membrane. Stimuli that can induce necroptosis include ligation of death receptors (a subset of the TNFR family), toll-like receptors (in particular, TLR3 and TLR4), interferons, and the intracellular viral sensor, DAI/ZBP1, among others. To study the process in more detail, it is useful to have a means to directly activate RIPK3. Here we provide protocols and procedures to artificially induce necroptotic cell death by drug-induced forced dimerization of RIPK3. We also provide information on specific kinase inhibitors, procedures to monitor RIPK3 and MLKL activation, and real-time quantification of cell death.

Key words Necroptosis, RIPK3, MLKL, Antibodies, Western blot, Cell death, Protein cross-linking

1 Introduction

The ability of RIPK3 to promote necroptosis can be modulated in both directions: induced by the activity of receptor interacting protein kinase-1 (RIPK1) [1] or antagonized by the proteolytic activity of a complex formed by RIPK1, FADD, caspase-8, and c-FLIP$_L$ [2–7]. RIPK3 serine/threonine kinase activation relies on RIP homotypic interaction motifs (RHIM). Hence, current evidence indicates that at least three important players, including RIPK1, TRIF, or ZBP1/DAI, can interact and activate RIPK3 via RHIM-RHIM interactions, thereby triggering necroptosis [1]. RIPK1 and RIPK3 form an amyloid-like signaling platform [8] but artificially enforced dimerization/oligomerization of RIPK3 (*see* Fig. 1) is sufficient to induce cell death in an MLKL-dependent manner [9–11]. RIPK3 autophosphorylation at Thr231 and Ser232 residues is essential to trigger necroptosis [12]. Inhibition of RIPK3 kinase activity by using small molecules is a viable approach for repressing necroptosis. Recently, a few effective

Adrian T. Ting (ed.), *Programmed Necrosis: Methods and Protocols*, Methods in Molecular Biology, vol. 1857, https://doi.org/10.1007/978-1-4939-8754-2_7,

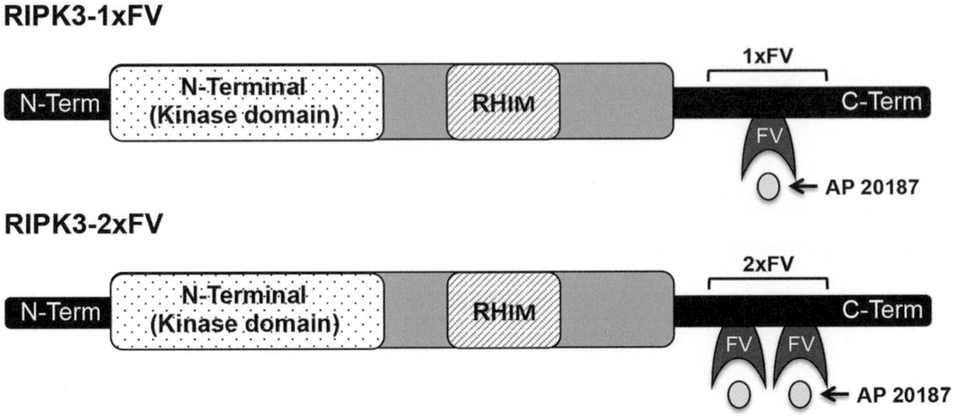

RIPK3-1xFV

1xFV
N-Term
N-Terminal (Kinase domain)
RHIM
C-Term
FV
AP 20187

RIPK3-2xFV

2xFV
N-Term
N-Terminal (Kinase domain)
RHIM
C-Term
FV
FV
AP 20187

Fig. 1 Primary structure of murine RIPK3 fused with one or two FV domains at the C-terminal. Representation illustrates two molecules of RIPK3 in presence of the dimerizer AP 20187 (gray circles)

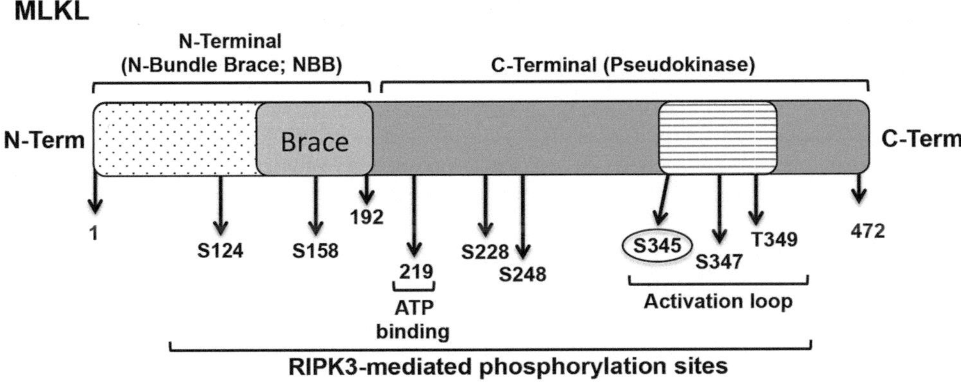

Fig. 2 Primary structure of murine MLKL. The N-terminal region containing N-Bundle Brace (NBB; residues 1–192) and C-terminal region (residues 193–472) containing the pseudokinase domain illustrate the most recently described RIPK3-mediated phosphorylation sites, including the Ser345 residue at the activation loop (circle)

inhibitors have been described for RIPK3, which include GW440139B (GW'39B), dabrafenib, GSK'843, and GSK'872 [13–15].

Direct binding and phosphorylation of MLKL by RIPK3 has been proposed to induce essential conformational changes in the "latch" of this pseudokinase, allowing the formation of oligomers [15], migration to the plasma membrane [16–19], and a sequential/hierarchical transduction of structural changes for the specific binding motifs to phosphatidylinositol lipids [20], directly disrupting the membranes' integrity [17, 21]. Recent studies described several sites of RIPK3-mediated phosphorylation of murine MLKL at the activation loop (Ser345, SerS347, and Thr349), and in the boundaries of the N-terminal domain (Ser158, Ser228, and Ser248), suggesting a fine-tuning for MLKL activity modulation [22, 23] (*see*

Fig. 2). Importantly, phosphorylation of Ser345 in murine MLKL is critical for RIPK3-mediated activation by either the TNF/TNR1 pathway or by forced dimerization of RIPK3 [15], indicating that Ser345 (Ser358 for human MLKL [24]) is a strong marker for RIPK3-induced MLKL activation (*see* Fig. 2). Here, we provide detailed information for inducing RIPK3-MLKL-mediated necroptotic cell death and the most relevant protocols and markers for detecting RIPK3 and MLKL phosphorylation and quantifying cell death. All methods and reagents described concern murine systems, although these are readily adapted for use in studies of human cells.

2 Materials

2.1 Reagents

All reagents should be dissolved in DMSO or water according to the manufacturer's instructions, aliquoted, and kept frozen at −20 °C.

1. Murine TNF (Peprotech).
2. RIPK1 Inhibitor II: 7-Cl-O-Nec-1 (Nec-1s) (Calbiochem).
3. zVAD-fmk (Apexbio).
4. Doxycycline (DOX) (Clontech).
5. Homodimerizer (AP 20187) (Clontech).
6. Cross-linker bismaleimidohexane (BMH) (ThermoFisher Scientific).
7. RIPK3 inhibitors: GSK2399872B (GSK'872) (Millipore). Alternatively, GW440139B (GW'39B) is not commercially available at present but can be obtained upon request from GlaxoSmithKline (GSK).
8. Sytox Green (Invitrogen/ThermoFisher Scientific).
9. Syto 24 Green (Invitrogen/ThermoFisher Scientific).
10. Propidium iodide (PI).
11. Annexin V-APC.
12. Retro-X Tet-On 3G Inducible Expression System (Clontech/Takara).

2.2 Cell Lines and Tissue Culture Media

1. Phoenix Amphotropic cells (Phoenix-AMPHO, ATCC® CRL-3213™).
2. Complete DMEM: DMEM (Life Technologies), 10% FBS, L-glutamine, pen/strep, 55 μM β-mercaptoethanol, 1 mM sodium pyruvate, and nonessential amino acids (Life Technologies).
3. Selection media: complete DMEM, 2 μg/mL puromycin, or alternatively, with 2 μg/mL puromycin plus 10 μg/mL blasticidin (Sigma-Aldrich) for the Tet-on system.

2.3 Antibodies

All solutions must be prepared by using ultrapure water and analytical grade reagents. Prepare and store all reagents at 4 °C (unless indicated otherwise).

1. RIPK3 antibodies: rabbit polyclonal R4277 (Sigma-Aldrich); rabbit polyclonal NBP-77299 (Novus).
2. Phosphorylated RIPK3 (Ser232) antibody: rabbit monoclonal clone EPR9516(N)-25 (Abcam).
3. MLKL antibodies: rabbit polyclonal ap14272b, epitope in C-terminal (Abgent); rat monoclonal clone 3H1, epitope in N-terminal, cross reacts with human MLKL (Millipore).
4. Phosphorylated MLKL (Ser345) antibodies: mouse monoclonal clone 7C6.1, (Millipore). This antibody has been tested in ELISA, IP, immunocytochemistry, and western blot (WB) assays [15]. Alternatively, another primary antibody (Abcam rabbit monoclonal clone EPR9515(2)) can be used for WB detection [25].

2.4 Western Blot and Cross-Linking

1. 1× RIPA buffer: 20 mM Tris–HCl pH 7.4, 150 mM NaCl, 1 mM Na_2EDTA, 1 mM EGTA, 1% NP-40, and 1% sodium deoxycholate and supplementing with protease and phosphatase inhibitors (Roche).
2. Lysis buffer for cross-linking: 50 mM Tris–HCl pH 7.4, 150 mM NaCl, 1 mM EDTA, and 0.5% NP-40 supplemented with protease and phosphatase inhibitors (Roche).
3. SDS-PAGE running buffer (Bio-Rad).
4. Transfer buffer: 25 mM Tris, 192 mM glycine, 20% methanol.
5. 10× TBS: 1.5 M NaCl, 0.1 M Tris–HCl, pH 7.4.
6. Criterion™ XT precast Gels, 4–12% (Bio-Rad).
7. Criterion™ Cell gel electrophoresis equipment (Bio-Rad).
8. Criterion™ Blotter gel transfer apparatus (Bio-Rad).
9. Nitrocellulose membrane (Bio-Rad).
10. Enhanced chemiluminescence kit (Bio-Rad).

2.5 Plasmids

1. Constructs for the production of retrovirus expressing fusion proteins of RIPK3 with one or two modified FKBP binding domains (1xFv and 2xFv, respectively) can be obtained by request from the authors, and are previously described [11] (*see* Fig. 1 and **Note 5**).
2. Constructs for the inducible expression of N-Terminal or C-Terminal MLKL can be obtained by request from the authors, and are previously described [13, 15].

3 Methods

3.1 Isolation, Culture, and Immortalization of Primary MLKL$^{-/-}$ or Ripk3$^{-/-}$ Mouse Embryonic Fibroblasts (MEFs)

1. After mating animals with the required genotypes, harvest embryos at E12–E14 (based on palpation) to generate primary MEFs.
2. Remove the head and fetal liver, and cut the remainder of the embryo in small pieces.
3. Disaggregate the small pieces in complete DMEM by passing them through a sterile syringe with an 18G needle, then pass the disaggregated tissues through a 40-μM sterile cell strainer into a 6-well plate.
4. Incubate for 2–3 days at 37 °C in the presence of 5% CO_2. Primary MEFs will attach, and other cell types will remain in suspension. After 3 days, pass the cells into a 10-cm dish and culture for another 4 days. To avoid overconfluence, do not pass more than 50% of the cells into a new 10-cm dish. (*see* **Note 1**).
5. After 2–3 passages, immortalize cells by infecting them with SV40 large T antigen [26]. Maintain cells in DMEM for 4 or 5 more passages (the immortalization process takes approximately 3–4 weeks).

3.2 Preparation of Cells for Western Blot Lysis and/or Cross-Linking Assays

1. For a typical 10-cm dish of cells treated to undergo necroptosis as indicated in Subheadings 3.6–3.8 below, wash plate once with 10 mL cold 1× PBS (*see* **Note 2**).
2. Harvest the cells in 10 mL of fresh cold PBS on ice using a cell scraper. Transfer the cells to a 15-mL tube and centrifuge them at 400 × *g* for 5 min. Discard supernatant.
3. The cell pellet can now be lysed for WB (*see* Subheadings 3.3 and 3.4) by resuspending pellet in 500 μL of 1× RIPA buffer on ice for 30 min. Alternatively, the pellet can be lysed for cross-linking assays (*see* Subheading 3.5).

3.3 Detection of MLKL or RIPK3

1. Load 10–30 μg of total proteins onto 4–12% SDS-PAGE precast gels and separate the proteins via electrophoresis. For Criterion™ XT precast gels, we typically run for 2 h at 100–120 volts.
2. Transfer the resolved protein onto nitrocellulose membranes for 90 min at 40 V.
3. Block membranes by incubating in 5% fat-free milk in 1× TBS for at least 1 h.
4. Prepare dilutions of 0.5–1.0 μg/mL of the primary antibodies for MLKL or RIPK3 in blocking buffer. If antibody stock concentration is at 1 mg/mL, this corresponds to a 1:1000–2000 dilution (*see* Figs. 3 and 4). Incubate membrane with primary antibody dilution for 4 h at room temperature or overnight with rotation at 4 °C.

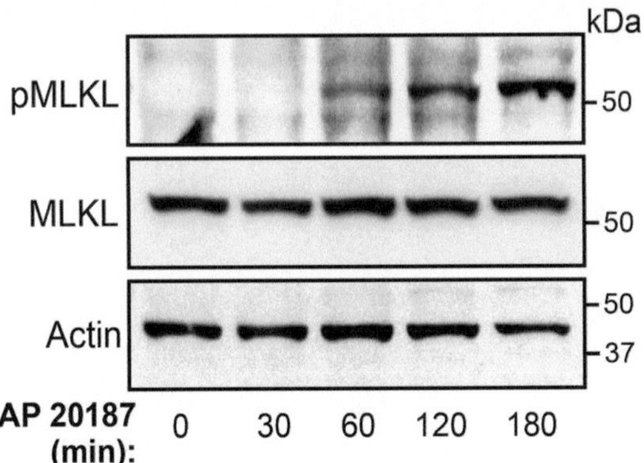

Fig. 3 Western blot analysis for pMLKL/total MLKL shows the effect of forced dimerization of RIPK3 in NIH-3T3 + RIPK3-2xFV incubated with 10 nM AP 20187 (modified from ref. 15)

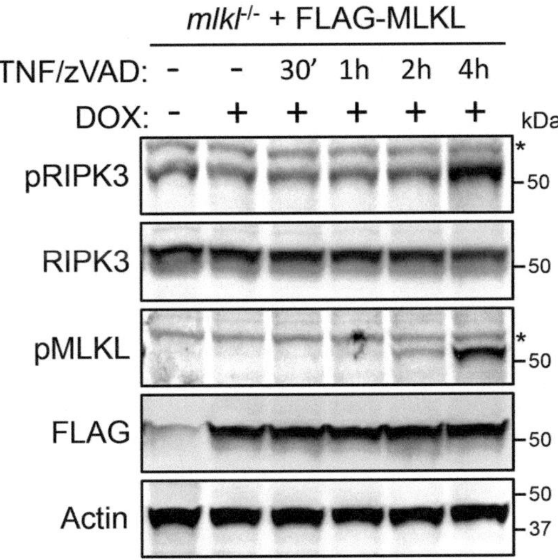

Fig. 4 Inducible Tet-on system for FLAG-MLKL (N-Term). *Mlkl*$^{-/-}$ + FLAG-MLKL cells were preincubated with DOX with a subsequent kinetic (at 0 min, 30 min, 1 h, 2 h, or 4 h) and treated with 10 ng/mL TNF plus 25 μM zVAD. Phosphorylated RIPK3 (Ser232), MLKL (Ser345), or both were detected as described in Subheadings 3.2 to 3.5. * Indicate nonspecific bands

5. Wash the nitrocellulose membrane three times (5–10 min) with 1% Tween in 1× TBS.
6. Incubate membrane at room temp with respective dilutions of HRP-conjugated secondary antibodies in 5% fat-free milk in 1× TBS (suggested 1:5000–10,000).
7. Wash three times (5–10 min) with 1% Tween in 1× TBS. The signal can be detected using an enhanced chemiluminescence kit.

3.4 Detection of Phosphorylated MLKL (pSer345-MLKL) or Phosphorylated RIPK3 (pSer232-RIPK3)

1. Electrophorese samples as indicated in Subheading 3.3, **step 1**.
2. Block membrane by incubating in 3% BSA in 1× TBS for at least 1 h at room temp.
3. Prepare dilutions of 0.5–1.0 μg/mL in blocking solution of the primary antibodies for pRIPK3 or pMLKL. Incubate overnight with agitation at 4 °C.
4. Proceed as indicated in **step 4** of Subheading 3.3 (*see* Figs. 3 and 4).

3.5 Detection of MLKL or RIPK3 Oligomerization in Cross-Linking Experiments

1. Collect cells treated to undergo necroptosis as described in Subheading 3.2. Lyse the cell pellet using 500 μL of the lysis buffer for cross-linking (*see* Subheading 2.4, **step 2**). Incubate the lysed cells for 30 min on ice or alternatively, on a plate rotator at 4 °C.
2. Centrifuge the cells at 400 × *g* for 5 min at 4 °C.
3. Transfer the supernatant to fresh tubes and quantify the total protein (*see* **Note 3**). Prepare duplicates of experimental tubes containing 0.5–1 mg of total protein in 200 μL (one set for cross-linking and the other set to monitor only monomers).
4. Prepare a stock solution of 25 mM BMH by dissolving 1.7 mg BMH into 244 μL DMSO (*see* **Note 4**), and subsequently make a 1:10 dilution of the stock BMH solution in DMSO to generate 2.5 mM BMH.
5. Add 5 μL of 2.5 mM BMH solution to ONLY one set of tubes containing 100 μL lysate to reach 125 μM BMH final concentration (*see* **Note 4**). Incubate at room temp for exactly 10 min and then stop the cross-linking reaction by adding 5 μL of 1 M DTT.
6. Finally, for ALL sets of tubes, add 30 μL of 4× SDS-PAGE sample buffer and boil 5 min at 90 °C. 20–40 μL of the samples can then be loaded onto a new gel for WB or stored at −80 °C (*see* Fig. 5).

3.6 Induction of Necroptosis by Forced Dimerization of RIPK3

3.6.1 Overexpression of Dimerizable RIPK3 in NIH-3T3 Cells or Ripk3$^{-/-}$ MEF

1. Transiently transfect Phoenix Amphotropic cells that have been previously plated in a 10-cm dish (desired confluence of 30–40%) with 4–6 μg of plasmid expressing RIPK3-1xFV or RIPK3-2xFV (*see* **Note 5**).
2. After 48 h, collect the supernatant, filter it through a 45-μm syringe filter, and use the filtered supernatant for the retroviral infection of *ripk3*$^{-/-}$ or NIH-3T3 cells (*see* **Note 6**).
3. After 72 h, *ripk3*$^{-/-}$ or NIH-3T3 cells expressing RIPK3-1xFV or RIPK3-2xFV should be selected by incubation with complete DMEM plus 2 μg/mL puromycin (*see* **Note 7**).

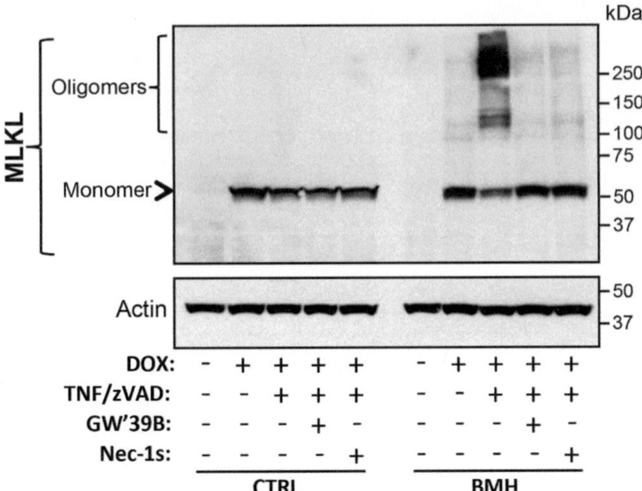

Fig. 5 *Mlkl*$^{-/-}$ MEF expressing FLAG-MLKL (N-Term) were cultured for 16 h with 1 μg/mL DOX followed by a 4-h treatment with 10 ng/mL TNF plus 25 μM zVAD in the absence or presence of 0.5 μM GW'39B or 30 μM Nec-1s. Cell lysates were incubated in the absence or presence of 125 μM BMH cross-linker, and the formation of MLKL oligomers was analyzed by western blot as indicated above (modified from ref. 15)

3.6.2 Forced Dimerization of RIPK3

1. After puromycin selection, plate cells into desired format for cell death analysis (*see* below) and WB analysis (*see* **Note 8**).
2. Perform a time course of 100 nM dimerizer (AP 20187) by incubating cells for 0, 30, 60, 120, and 180 min (*see* example at Fig. 3 and **Note 9**) and also titration of AP 20187 as described [11] (for more details for RIPK3-1xFV versus RIPK3-2xFV *see* **Note 9**).
3. After treatment, collect cells for WB and/or FACS analysis [15]. Phosphorylation of RIPK3 and/or MLKL can be monitored by WB as indicated in Subheadings 3.2–3.4. Alternatively, MLKL or RIPK3 oligomerization can be visualized by using lysis buffer and following the protocol described in Subheading 3.5. Forced dimerization of RIPK3 can be applied for either mouse (*see* Fig. 1) or human studies (*see* more details in ref. 20) [11, 15, 20].

3.6.3 Real-Time Quantification of Necroptotic Cell Death and Protection by Inhibiting RIPK's Kinase Activity

Important: This protocol requires an IncuCyte ZOOM imaging system (Essen Bioscience), which enables monitoring and quantification of cell death in real time. Other live cell imaging systems can be adapted to this protocol, depending on the parameters of the system.

1. To monitor RIPK3-induced necroptosis, plate cells in at least triplicate (*see* **Note 10**) according to desired experimental conditions (e.g., Control, 1, 10, and 100 nM AP 20187).

RIPK3-mediated necroptosis can be inhibited by adding RIPK3 kinase inhibitors (0.5–1 μM GSK'872 or 0.25–0.5 μM GW'39B).

2. To monitor dead and permeable cells, add all combinations of drug plus medium containing 25 nM of the membrane-impermeable dye Sytox Green. At the endpoint, stain all cells with 100 nM of the membrane-permeable dye Syto 24 Green (*see* **Note 11**).

3.6.4 FACS Analysis

Alternatively, cell death can be monitored by flow cytometry analysis by staining cells with AnnexinV-APC and propidium iodide (PI) before flow cytometry. Cell death can be expressed as the percentage calculated by sum of the percentage of AnnexinV$^+$PI$^+$ cells and the percentage of AnnexinV$^+$PI$^-$ cells. In this case, plate cells as indicated above, and treat the cells until the time point desired is reached.

1. Collect the cells by trypsinization and then resuspend them in 1× PBS containing 1 μg/mL PI.
2. Perform subsequent staining with AnnexinV-APC (see additional information in ref. 15).

3.7 Inducible Tet-On System for Mouse MLKL

Reconstitution of mouse MLKL FLAG-tagged at the N- or C-terminal: Two major steps are required to generate this inducible system and reconstitute MLKL into *mlkl*$^{-/-}$ (or any target murine cells): step A is to introduce the "Regulator" (pRetroX-Tet3G), and step B is to introduce FLAG-MLKL (N-Term) or MLKL-FLAG (C-Term) (previously reported in [13]) cloned into the doxycycline-inducible vector pRetroX-TRE3G.

1. The retroviruses for steps A and B are generated as described in Subheading 3.6.1.
2. After proper selection in media containing 10 mg/mL blasticidin and 2 μg/mL puromycin, *mlkl*$^{-/-}$ (or any cell line being used) reconstituted with either FLAG-MLKL (N-Term) or MLKL-FLAG (C-Term) is ready for doxycycline induction of MLKL expression (*see* **Note 12**). As observed in Fig. 4, *mlkl*$^{-/-}$ MEFs + FLAG-MLKL have detectable MLKL oligomers (analyzed as described in Subheadings 3.2 to 3.5) after necroptotic signal induction.

3.8 Upstream Activation of the RIPK3-MLKL Pathway

All systems described above, which includes WT, mlkl$^{-/-}$ + MLKL-FLAG or ripk3$^{-/-}$ plus RIPK3-2xFV MEFs, are suitable for studying necroptosis induced by TNF/TNFR1 pathway ([9, 13, 15]).

1. Plate cells as described in Subheading 3.6.2 (*see* **Note 4**). After 24 h, incubate the cells for 0, 1, 2, or 4 h in different plates with complete DMEM containing one of the following: DMSO (control); 10 ng/mL TNF, 25 μM zVAD, or 10 ng/mL

TNF plus 25 μM zVAD (*see* **Note 13**). Necroptosis induced by TNF plus zVAD can be prevented by combining with 15–30 μM Nec-1s, 0.25–0.5 μM GW'39'B, or 0.5–1.0 μM GSK'872.

2. Proceed as indicated in Subheadings 3.2–3.5 (*see* **Note 14**).

4 Notes

1. Different types of primary cells have different replicative rates of division, which strongly depend on the phenotype of the genes; consequently, the times indicated can vary between cell types.
2. This protocol is recommended for adherent cells that are well attached to the plates. Alternatively, cells can be harvested by using a cell scraper directly on cells in culture media (always on ice), then centrifuging the harvested cells at 400 × *g* at 4 °C for 5 min and washing the cell pellet twice with cold PBS.
3. It is desirable to have a concentration of total proteins that reaches 0.5–1 mg in a volume of 200 μL for the cross-linking reaction. A lower amount of total proteins will not produce a good signal for detection of oligomers of MLKL.
4. BMH (bismaleimidohexane) must be stored protected from light and at 4 °C in a desiccator. Always prepare fresh solutions for each experiment, and protect them from light. BMH is very unstable, but fresh solutions will provide reproducible results. Always work quickly and precisely with the incubation times. The final working concentration depends on the abundance of the protein of interest. For murine MLKL, a proper range is 125–250 μM.
5. The plasmid constructs expressing 2xFV RIPK3 (N-Term or C-Term) do not affect RIPK3 kinase activity or interaction with MLKL: however, for other proteins, the tag on the N- or C-terminal domain could be a determinant of the activity/function. The plasmids expressing RIPK3-1xFV, RIPK3-2xFV, RIPK3$^{\Delta RHIM}$-1xFV or RIPK3$^{\Delta RHIM}$-2xFV were previously described [9, 11].
6. To improve the efficacy of retroviral infections, start with low confluence of both cell systems (Phoenix-AMPHO, ATCC® CRL-3213™, and NIH-3T3). Also, it is recommended to perform two consecutive retroviral infections by collecting additional supernatant from AMPHO cells after 72 h.
7. NIH-3T3 cells expressing RIPK3-2xFV or RIPK3$^{\Delta RHIM}$-2xFV with or without an amino-terminal (N-Term) FLAG-tag were previously described [9, 11]. This protocol can be applied to any murine cell line in the absence or presence of

endogenous RIPK3 or MLKL. The efficacy of selection of cells overexpressing RIPK3 strongly depends on the efficiency of retroviral infection. It is important to incorporate the proper controls for M.O.I. (e.g., by infecting with reporter genes such as GFP).

8. Detection of pMLKL/MLKL by western blotting requires high endogenous expression levels. To increase pMLKL detection, a larger format should be used for total protein extraction (i.e., 10-cm dish). Alternatively, endogenous MLKL levels can be increased by priming the cells with low amounts of IFNβ (25 U/mL) [15]. Cell death experiments can be performed by using 6-, 12-, 24-, or 48-well plates with a cell density of 2.5×10^6 cells/plate on the day of the experiment.
9. Titrating AP 20187 is a good complementary experiment in order to set up the proper concentration of required dimerizer to force RIPK3 activation. In most cases, a proper AP 20187 concentration is between 10–100 nM. Of note, forced dimerization of RIPK3-1xFV is regulated by RIPK1 and caspase 8, and consequently, zVAD enhances RIPK3 activation and Nec-1s can repress it. However, AP-1-induced oligomerization of RIPK3-2xFV overcomes that modulation and RIPK3 activation cannot be repressed by Nec-1s [11].
10. Cell density is important at this step and strongly depends on cell type. For example, for MEF cells, a proper density requires seeding 1.4×10^6 cells per plate (6-, 12-, 24-, or 48-well plates) the day before treatment.
11. Experimental data can be expressed as percentages of the ratio between Sytox Green/Syto 24 (% Sytox Green+) or expressed as the absolute number of Sytox Green-positive events per well (IncuCyte image analysis software [Essen Bioscience]).
12. For more details, consult manufacturer instructions (Retro-XTM Tet-On® 3G inducible expression system, Clontech, Cat No. 631188). For MLKL-inducible (N- or C-Term FLAG MLKL) systems, consult refs. 13, 15.
13. The inducible system for MLKL requires addition of 1 μg/mL of DOX; this can be aggregated with the other drugs, or cells can be preincubated overnight in DOX (~16 h) and then incubated under necroptotic conditions (this provides fast cell-death kinetics and activation of RIPK1/RIPK3/MLKL). We suggest using a dose of 10 ng/mL TNF; however, depending on the cell types used, sensitivity to TNF can vary, so titration of TNF and zVAD is recommended.
14. The interaction of necrosome components can be monitored by immunoprecipitation using most of the experimental settings presented here (*see* more details in refs. 13, 15).

References

1. Li J, McQuade T, Siemer AB, Napetschnig J, Moriwaki K, Hsiao YS, Damko E, Moquin D, Walz T, McDermott A, Chan FK, Wu H (2012) The RIP1/RIP3 necrosome forms a functional amyloid signaling complex required for programmed necrosis. Cell 150(2):339–350. https://doi.org/10.1016/j.cell.2012.06.019
2. Green DR, Oberst A, Dillon CP, Weinlich R, Salvesen GS (2011) RIPK-dependent necrosis and its regulation by caspases: a mystery in five acts. Mol Cell 44(1):9–16. https://doi.org/10.1016/j.molcel.2011.09.003
3. Weinlich R, Oberst A, Dillon CP, Janke LJ, Milasta S, Lukens JR, Rodriguez DA, Gurung P, Savage C, Kanneganti TD, Green DR (2013) Protective roles for caspase-8 and cFLIP in adult homeostasis. Cell Rep 5(2):340–348. https://doi.org/10.1016/j.celrep.2013.08.045
4. Dillon CP, Oberst A, Weinlich R, Janke LJ, Kang TB, Ben-Moshe T, Mak TW, Wallach D, Green DR (2012) Survival function of the FADD-CASPASE-8-cFLIP(L) complex. Cell Rep 1(5):401–407. https://doi.org/10.1016/j.celrep.2012.03.010
5. Oberst A, Dillon CP, Weinlich R, McCormick LL, Fitzgerald P, Pop C, Hakem R, Salvesen GS, Green DR (2011) Catalytic activity of the caspase-8-FLIP(L) complex inhibits RIPK3-dependent necrosis. Nature 471(7338):363–367. https://doi.org/10.1038/nature09852
6. Kaiser WJ, Upton JW, Long AB, Livingston-Rosanoff D, Daley-Bauer LP, Hakem R, Caspary T, Mocarski ES (2011) RIP3 mediates the embryonic lethality of caspase-8-deficient mice. Nature 471(7338):368–372. https://doi.org/10.1038/nature09857
7. Zhang H, Zhou X, McQuade T, Li J, Chan FK, Zhang J (2011) Functional complementation between FADD and RIP1 in embryos and lymphocytes. Nature 471(7338):373–376. https://doi.org/10.1038/nature09878
8. Wu XN, Yang ZH, Wang XK, Zhang Y, Wan H, Song Y, Chen X, Shao J, Han J (2014) Distinct roles of RIP1-RIP3 hetero- and RIP3-RIP3 homo-interaction in mediating necroptosis. Cell Death Differ 21(11):1709–1720. https://doi.org/10.1038/cdd.2014.77
9. Tait SW, Oberst A, Quarato G, Milasta S, Haller M, Wang R, Karvela M, Ichim G, Yatim N, Albert ML, Kidd G, Wakefield R, Frase S, Krautwald S, Linkermann A, Green DR (2013) Widespread mitochondrial depletion via mitophagy does not compromise necroptosis. Cell Rep 5(4):878–885. https://doi.org/10.1016/j.celrep.2013.10.034
10. Cook WD, Moujalled DM, Ralph TJ, Lock P, Young SN, Murphy JM, Vaux DL (2014) RIPK1- and RIPK3-induced cell death mode is determined by target availability. Cell Death Differ 21(10):1600–1612. https://doi.org/10.1038/cdd.2014.70
11. Orozco S, Yatim N, Werner MR, Tran H, Gunja SY, Tait SW, Albert ML, Green DR, Oberst A (2014) RIPK1 both positively and negatively regulates RIPK3 oligomerization and necroptosis. Cell Death Differ 21(10):1511–1521. https://doi.org/10.1038/cdd.2014.76
12. Chen W, Zhou Z, Li L, Zhong CQ, Zheng X, Wu X, Zhang Y, Ma H, Huang D, Li W, Xia Z, Han J (2013) Diverse sequence determinants control human and mouse receptor interacting protein 3 (RIP3) and mixed lineage kinase domain-like (MLKL) interaction in necroptotic signaling. J Biol Chem 288(23):16247–16261. https://doi.org/10.1074/jbc.M112.435545
13. Dillon CP, Weinlich R, Rodriguez DA, Cripps JG, Quarato G, Gurung P, Verbist KC, Brewer TL, Llambi F, Gong YN, Janke LJ, Kelliher MA, Kanneganti TD, Green DR (2014) RIPK1 blocks early postnatal lethality mediated by caspase-8 and RIPK3. Cell 157(5):1189–1202. https://doi.org/10.1016/j.cell.2014.04.018
14. Kaiser WJ, Sridharan H, Huang C, Mandal P, Upton JW, Gough PJ, Sehon CA, Marquis RW, Bertin J, Mocarski ES (2013) Toll-like receptor 3-mediated necrosis via TRIF, RIP3, and MLKL. J Biol Chem 288(43):31268–31279. https://doi.org/10.1074/jbc.M113.462341
15. Rodriguez DA, Weinlich R, Brown S, Guy C, Fitzgerald P, Dillon CP, Oberst A, Quarato G, Low J, Cripps JG, Chen T, Green DR (2016) Characterization of RIPK3-mediated phosphorylation of the activation loop of MLKL during necroptosis. Cell Death Differ 23(1):76–88. https://doi.org/10.1038/cdd.2015.70
16. Cai Z, Jitkaew S, Zhao J, Chiang HC, Choksi S, Liu J, Ward Y, Wu LG, Liu ZG (2014) Plasma membrane translocation of trimerized MLKL protein is required for TNF-induced necroptosis. Nat Cell Biol 16(1):55–65. https://doi.org/10.1038/ncb2883
17. Wang H, Sun L, Su L, Rizo J, Liu L, Wang LF, Wang FS, Wang X (2014) Mixed lineage kinase domain-like protein MLKL causes necrotic membrane disruption upon phosphorylation by RIP3. Mol Cell 54(1):133–146. https://doi.org/10.1016/j.molcel.2014.03.003
18. Hildebrand JM, Tanzer MC, Lucet IS, Young SN, Spall SK, Sharma P, Pierotti C, Garnier JM, Dobson RC, Webb AI, Tripaydonis A, Babon JJ, Mulcair MD, Scanlon MJ, Alexander WS, Wilks AF, Czabotar PE, Lessene G, Murphy

JM, Silke J (2014) Activation of the pseudokinase MLKL unleashes the four-helix bundle domain to induce membrane localization and necroptotic cell death. Proc Natl Acad Sci U S A 111(42):15072–15077. https://doi.org/10.1073/pnas.1408987111

19. Chen X, Li W, Ren J, Huang D, He WT, Song Y, Yang C, Zheng X, Chen P, Han J (2014) Translocation of mixed lineage kinase domain-like protein to plasma membrane leads to necrotic cell death. Cell Res 24(1):105–121. https://doi.org/10.1038/cr.2013.171
20. Quarato G, Guy CS, Grace CR, Llambi F, Nourse A, Rodriguez DA, Wakefield R, Frase S, Moldoveanu T, Green DR (2016) Sequential engagement of distinct MLKL phosphatidylinositol-binding sites executes necroptosis. Mol Cell 61(4):589–601. https://doi.org/10.1016/j.molcel.2016.01.011
21. Dondelinger Y, Declercq W, Montessuit S, Roelandt R, Goncalves A, Bruggeman I, Hulpiau P, Weber K, Sehon CA, Marquis RW, Bertin J, Gough PJ, Savvides S, Martinou JC, Bertrand MJ, Vandenabeele P (2014) MLKL compromises plasma membrane integrity by binding to phosphatidylinositol phosphates. Cell Rep 7(4):971–981. https://doi.org/10.1016/j.celrep.2014.04.026
22. Murphy JM, Czabotar PE, Hildebrand JM, Lucet IS, Zhang JG, Alvarez-Diaz S, Lewis R, Lalaoui N, Metcalf D, Webb AI, Young SN, Varghese LN, Tannahill GM, Hatchell EC, Majewski IJ, Okamoto T, Dobson RC, Hilton DJ, Babon JJ, Nicola NA, Strasser A, Silke J, Alexander WS (2013) The pseudokinase MLKL mediates necroptosis via a molecular switch mechanism. Immunity 39(3):443–453. https://doi.org/10.1016/j.immuni.2013.06.018
23. Tanzer MC, Tripaydonis A, Webb AI, Young SN, Varghese LN, Hall C, Alexander WS, Hildebrand JM, Silke J, Murphy JM (2015) Necroptosis signalling is tuned by phosphorylation of MLKL residues outside the pseudokinase domain activation loop. Biochem J 471(2):255–265. https://doi.org/10.1042/BJ20150678
24. Sun L, Wang H, Wang Z, He S, Chen S, Liao D, Wang L, Yan J, Liu W, Lei X, Wang X (2012) Mixed lineage kinase domain-like protein mediates necrosis signaling downstream of RIP3 kinase. Cell 148(1-2):213–227. https://doi.org/10.1016/j.cell.2011.11.031
25. Cai Z, Zhang A, Choksi S, Li W, Li T, Zhang XM, Liu ZG (2016) Activation of cell-surface proteases promotes necroptosis, inflammation and cell migration. Cell Res 26(8):886–900. https://doi.org/10.1038/cr.2016.87
26. Conzen SD, Cole CN (1995) The three transforming regions of SV40 T antigen are required for immortalization of primary mouse embryo fibroblasts. Oncogene 11(11):2295–2302

Chapter 8

Detection of MLKL Oligomerization During Programmed Necrosis

Zhenyu Cai and Zheng-Gang Liu

Abstract

Programmed necrosis, also known as necroptosis, is a form of regulated necrotic cell death that is mediated by receptor-interacting protein kinases RIP1 (or RIPK1), RIP3 (or RIPK3), and the mixed lineage kinase domain-like protein, MLKL. Following the induction of programmed necrosis, MLKL is phosphorylated by RIP3 and oligomerizes and then the protein translocates to cell plasma membrane in order to execute programmed necrosis. Here, we describe a detailed protocol to detect MLKL oligomerization in necroptotic cells by Western blotting analysis under nonreducing condition. Therefore, we established the method to detect the activation of programmed necrotic pathway.

Key words MLKL, Necroptosis, Nonreducing SDS-PAGE, Immunoblots

1 Introduction

Programmed necrosis, also known as necroptosis, is a unique regulated mode of necrosis, characterized by the rounding of cell shape, the increase of cell volume, the rupture of plasma membrane and finally, the release of intracellular contents [1, 2]. Distinct from nonregulated necrosis, programmed necrosis is mediated by the core death machinery that consists of the receptor-interacting protein kinases RIP1 (or RIPK1), RIP3 (or RIPK3) and the mixed lineage kinase domain-like, MLKL [3–8]. In cells undergoing programmed necrosis in response to different stimuli such as tumor necrosis factor (TNF) plus Smac mimetic and the caspase inhibitor z-VAD-fmk (TSZ), RIP1 phosphorylates and activates RIP3 and in turn, the activated RIP3 recruits MLKL to form the necrosome, in which RIP3 phosphorylates MLKL [4, 9]. The phosphorylated MLKL then oligomerizes and translocates to cell plasma membrane to execute the death process [10–13]. Our previous study suggested that the disulfide bonds of the oligomerized MLKL proteins formed by oxidation during cell lysis and that the cross-linked MLKL oligomer could be detected with anti-MLKL antibody by

Adrian T. Ting (ed.), *Programmed Necrosis: Methods and Protocols*, Methods in Molecular Biology, vol. 1857, https://doi.org/10.1007/978-1-4939-8754-2_8,

Western blotting under nonreducing condition [10]. Here we describe a detailed method to detect oligomeric MLKL by Western blotting analysis in programmed necrotic cells.

2 Materials

All solutions are prepared with ultrapure water and analytical grade reagents.

2.1 SDS Polyacrylamide Gel

1. Separation gel buffer: 1.5 M Tris–HCl Buffer, pH 8.8. Dissolve 181.65 g Tris base in around 700 mL of ddH_2O. Adjust the pH to 8.8 with concentrated HCl. Bring up the volume to 1 L with ddH_2O. Store at 4 °C (*see* **Note 1**).
2. Stacking gel buffer: 0.5 M Tris–HCl Buffer, pH 6.8. Dissolve 60.6 g Tris base in 700 mL distilled water. Adjust the pH to 6.8 with HCl and make up the volume to 1 L. Store at 4 °C.
3. 30% acrylamide/Bis solution (29.2:0.8): Dissolve 290 g of acrylamide and 10 g of N,N′-methylbisacrylamide in 600 mL of H_2O (*see* **Note 2**). Sterilize the solution by filtration through passing a filter with 0.45 micron pore size. Check the pH (should be 7.0 or less) and store at 4 °C, in a bottle wrapped with aluminum foil.
4. Ammonium persulfate: 10% solution in water.
5. N,N,N,N′-Tetramethylethylenediamine (TEMED): Store at 4 °C.
6. 10% separation gel (10 mL): 3.8 mL H_2O; 3.4 mL 30% acrylamide/Bis solution; 2.6 mL separation gel buffer; 100 μL 10% SDS; 100 μL 10% ammonium persulfate; 10 μL TEMED.
7. 5% stacking gel (5 mL): 3.4 mL H_2O; 0.83 mL 30% acrylamide/Bis solution; 0.63 mL stacking gel buffer; 50 μL 10% SDS; 50 μL 10% ammonium persulfate; 5 μL TEMED.
8. SDS-PAGE running buffer: 25 mM Tris, pH 8.3, 0.192 M glycine, 0.1% SDS.
9. You can also use precast gels from other commercial companies, for example, Criterion™ TGX™ Precast Gels (Bio-Rad).

2.2 Immunoblotting

1. PVDF membranes (Immobilon-P Membrane, PVDF membrane, 0.45 μm [Millipore]).
2. Transfer buffer: 25 mM Tris, 190 mM glycine, 20% methanol. Check the pH and adjust to pH 8.3 if necessary.
3. PBS buffer containing 0.05% Tween-20 (PBST). Dissolve the following in 800 mL distilled H_2O. 8 g of NaCl; 0.2 g of KCl; 1.44 g of Na_2HPO4; 0.24 g of KH_2PO4. Adjust the pH to 7.4

with HCl and bring up the volume to 1 L with ddH_2O. Then add 0.5 mL Tween-20 to 1 L 1× PBS.

4. Blocking buffer: 5% fat-free milk in PBST.
5. Antibody dilution buffer: 5% fat-free milk in PBST.
6. Criterion™ Blotter (Bio-Rad).
7. Plastic container.
8. Thick Blot Filter Paper (Bio-Rad).

2.3 Sample Preparation

1. M2 cell lysis buffer: 20 mM Tris, pH 7, 0.5% NP40, 250 mM NaCl, 3 mM EDTA, 3 mM EGTA, 2 mM DTT, 0.5 mM PMSF, 20 mM glycerol phosphate, 1 mM sodium vanadate, 1 μg/mL aprotinin and 1 μg/mL leupeptin.
2. 2× Laemmli loading buffer: 4% SDS, 20% glycerol, 0.004% bromophenol blue, 0.125 M Tris–HCl, Check the pH and bring it to pH 6.8 (*see* **Note 3**).

3 Antibody and Protein Marker

1. Anti-human MLKL antibody clone EPR17514 (Abcam).
2. Anti-mouse MLKL antibody was a gift from Jiahuai Han (Xiamen University, Xiamen, China) [11].
3. Prestained protein standard (Bio-Rad).

4 Reagents for Inducing Programmed Necrosis

1. Human TNFα (R&D System) is dissolved in ultrapure water at stock concentration of 10 ng/μL.
2. Mouse TNFα (R&D System) is dissolved in ultrapure water at stock concentration of 10 ng/μL.
3. Smac mimetic (SM-164) was a gift from Shaomeng Wang (University of Michigan, Ann Arbor, MI, USA) [14].
4. z-Val-Ala-Asp-(OMe)-Fluoromethyl Ketone (z-VAD-FMK) (R&D System) is dissolved in DMSO at stock concentration of 20 mM.

5 Methods

Carry out all procedures at room temperature, unless otherwise specified.

5.1 Cell Lysis

1. To induce programmed necrosis, HT29, Jurkat, or U937 human cells are pretreated with z-VAD-fmk (20 μM) and Smac mimetic (10 nM) for 30 min and are then treated with human TNFα (30 ng/mL) for 4 h. To induce programmed necrosis in mouse embryonic fibroblasts cells (MEFs), the cells are pretreated with z-VAD-fmk (20 μM) and Smac mimetic (10 nM) for 30 min and then with mouse TNFα (15 ng/mL) for 4 h. For mouse L929 cells, they are pretreated with z-VAD-fmk (20 μM) for 30 min and then with mouse TNFα (15 ng/mL) for 2 h. Before being collected, these cells are washed with ice-cold PBS twice and are lysed in ice-cold M2 buffer (1 mL per 10^7 cells/100 mm dish/150 cm^2 flask; 0.5 mL per 5×10^6 cells/60 mm dish/75 cm^2 flask).
2. Use a cold plastic cell scraper to scrape adherent cells off from dishes and then gently transfer the cell suspensions into precooled microcentrifuge tubes and put tubes on ice.
3. Put the microcentrifuge tubes in a rotator to incubate at 4 °C for 30 min.
4. Centrifuge (12,000 rpm/13,000 × *g*) in a microcentrifuge at 4 °C for 5 min.
5. Gently remove the tubes from the centrifuge and place them on ice, transfer the supernatants into new tubes and keep them on ice. Discard the pellets.
6. To determine protein concentrations of these samples, we use a bicinchoninic acid (BCA) assay. Bovine serum albumin (BSA) is used as protein standard.
7. Once the concentrations of these samples are determined, they can be kept at −20 °C or − 80 °C for later use, if not immediately used for Western blotting.

5.2 Sample Preparation for Western Blotting

1. Based on the protein concentration of each sample, take 30 μg of total proteins for each sample and add 2× Laemmli sample buffer. Important: No DTT or β-mercaptoethano in sample buffer in order to keep the proteins under nonreducing condition (*see* **Note 3**).
2. Then boil each sample at 95 °C for 5 min. Lysates can be stored at −20 °C for future use, if not used immediately for Western blotting.

5.3 Load and Run the SDS-PAGE Gel

1. Load the samples into the wells of the SDS-PAGE gel along with molecular weight marker.
2. Run the gel for 1.5 h. at 100 V.

5.4 Transfer Proteins from the Gel to PVDF Membrane

1. Cut PVDF membrane and filter paper to the dimensions of the gel.
2. Treat PVDF membrane with methanol for 1 min and rinse it with transfer buffer before setting up the transfer stack.
3. Equilibrate the gel and soak the filter paper and fiber pads in transfer buffer for 5 min.
4. Use the transfer cassette of the wet transfer apparatus (Bio-Rad) to set up the transfer stack.
5. Place one presoaked fiber pad on the black side of the cassette.
6. Then place two sheets of presoaked filter paper over the fiber pad.
7. Next transfer the equilibrated gel onto the filter paper.
8. Place the pretreated PVDF membrane over the gel.
9. Complete the stack by placing two pieces of filter paper over the PVDF membrane and then a fiber pad.
10. To be sure that there is no air bubble between filter paper and the gel or the gel and the PVDF membrane. Using a glass pipette to gently roll air bubbles out.
11. Close the cassette firmly and be careful not to move the gel and filter paper stack. Lock the cassette with the white latch.
12. Place the cassette in the transfer tank and fill the rank with transfer buffer.
13. The time and voltage of transfer may require some optimization. We usually use 100 V for 1 h.
14. After the transfer, disassemble the stack to get the PVDF membrane. Mark the membrane on the side that faces the gel and the position of the pre-stained markers, since they may fade away during antibody incubation.

5.5 Detecting MLKL Oligomer with Anti-MLKL Antibody

1. Block the PVDF membrane with blocking buffer for 1 h.
2. Dilute the MLKL antibody (1: 2000) with antibody dilution buffer.
3. Incubate the PVDF membrane with the diluted MLKL antibody on a shaker at 4 °C overnight.
4. Wash the blot with PBST for 5 min, repeat three times.
5. Incubate the blot with the secondary antibody (peroxidase-conjugated goat anti-rabbit IgG, etc.) at room temperature for 1 h.
6. Wash the blot with PBST for 5 min, repeat three times.
7. To detect the MLKL protein with ECL (Thermo Scientific), follow the kit manufacturer's recommendations. Remove excess reagent and cover the membrane in transparent plastic wrap.

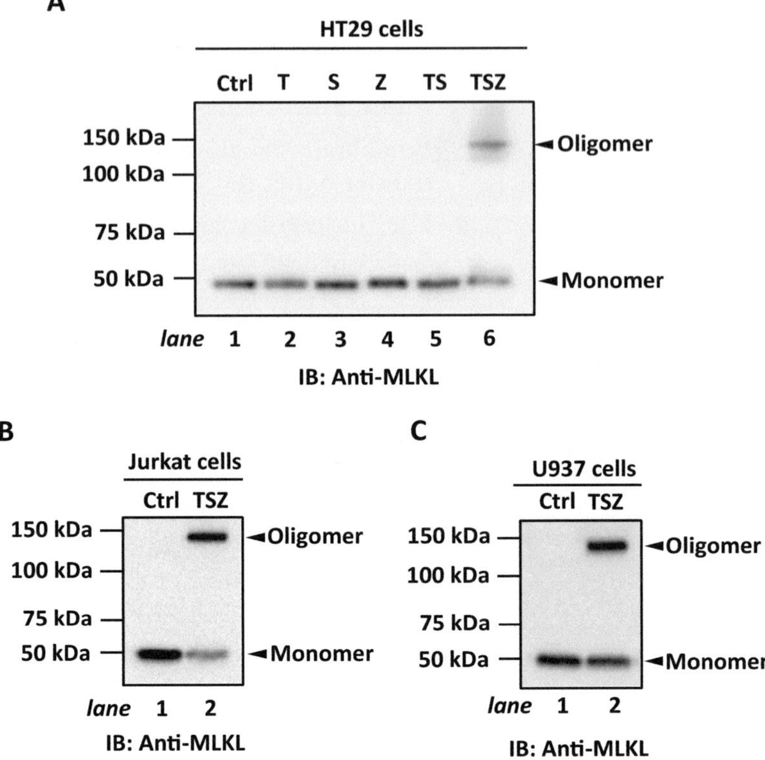

Fig. 1 Detection of MLKL oligomer in human necroptotic cells. (**a**) HT29 cells were treated with TNF (T, 30 ng/mL), Smac mimetic (S, 10 nM), Z-VAD-FMK (Z, 20 nM), TNF plus Smac mimetic (TS) to induce apoptosis or TNF plus Smac mimetic and Z-VAD-FMK (TSZ) to induce necroptosis for 4 h, respectively. The cell lysates were resolved on nonreducing gel and analyzed by immunoblotting with anti-MLKL antibody. ***Lane 6*** shows that the MLKL oligomer that can only be detected in necroptotic HT29 cells. (**b**) Jurkat or (C) U937 cells were treated with TSZ for 4 h. The cell lysates were resolved on nonreducing gel and analyzed by immunoblotting with anti-MLKL antibody. ***Lane 2*** shows that the MLKL oligomer that can only be detected in TSZ-induced necroptotic cells

8. To acquire the results of MLKL protein detection, place X-ray film over the blot for adequate exposure depending on the strength of the signal and develop the film. Representative results of MLKL oligomers detection are shown in Figs. 1 and 2

6 Notes

1. Make sure to let the solution cool down to room temperature before making the final pH adjustment.
2. Heating may be necessary to dissolve the acrylamide. Bring up the volume to 1 L with H_2O.

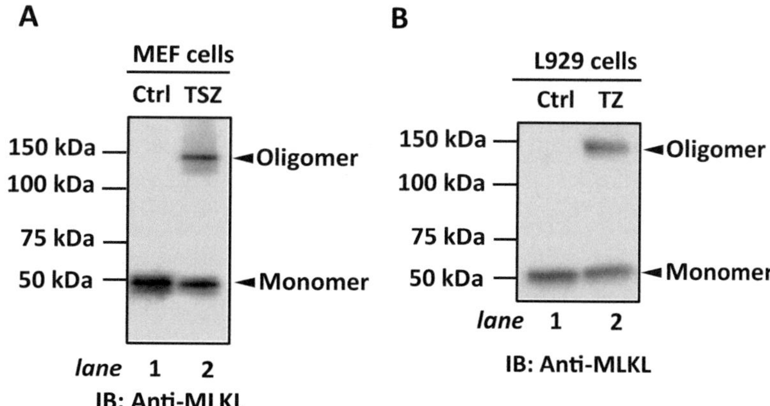

Fig. 2 Detection of MLKL oligomer in mouse necroptotic cells. (**a**) Mouse embryonic fibroblasts cells (MEFs) were treated with TSZ for 4 h. The cell lysates were resolved on nonreducing gel and analyzed by immunoblotting with anti-MLKL antibody. ***Lane 2*** shows that the MLKL oligomer can only be detected in necroptotic MEF cells. (**b**) Mouse L929 cells were treated with TZ for 2 h. The cell lysates were resolved on nonreducing gel and analyzed by immunoblotting with anti-MLKL antibody. ***Lane 2*** shows that the MLKL oligomer can only be detected in necroptotic L929 cells

3. In order to provide a nonreducing condition, M2 cell lysis buffer and 2× laemmli loading buffer do not contain any reducing agent such as DTT or 2-mercaptoethanol.

Acknowledgments

This research was supported by the Intramural Research Program of the Center for Cancer Research, National Cancer Institute, National Institutes of Health.

References

1. Zong WX, Thompson CB (2006) Necrotic death as a cell fate. Genes Dev 20(1):1–15. https://doi.org/10.1101/gad.1376506
2. Chan FK, Luz NF, Moriwaki K (2015) Programmed necrosis in the cross talk of cell death and inflammation. Annu Rev Immunol 33:79–106. https://doi.org/10.1146/annurev-immunol-032414-112248
3. Zhang DW, Shao J, Lin J, Zhang N, Lu BJ, Lin SC, Dong MQ, Han J (2009) RIP3, an energy metabolism regulator that switches TNF-induced cell death from apoptosis to necrosis. Science 325(5938):332–336. https://doi.org/10.1126/science.1172308
4. Sun L, Wang H, Wang Z, He S, Chen S, Liao D, Wang L, Yan J, Liu W, Lei X, Wang X (2012) Mixed lineage kinase domain-like protein mediates necrosis signaling downstream of RIP3 kinase. Cell 148(1-2):213–227. https://doi.org/10.1016/j.cell.2011.11.031
5. Holler N, Zaru R, Micheau O, Thome M, Attinger A, Valitutti S, Bodmer JL, Schneider P, Seed B, Tschopp J (2000) Fas triggers an alternative, caspase-8-independent cell death pathway using the kinase RIP as effector molecule. Nat Immunol 1(6):489–495
6. Lin Y, Choksi S, Shen HM, Yang QF, Hur GM, Kim YS, Tran JH, Nedospasov SA, Liu

ZG (2004) Tumor necrosis factor-induced nonapoptotic cell death requires receptor-interacting protein-mediated cellular reactive oxygen species accumulation. J Biol Chem 279(11):10822–10828. https://doi.org/10.1074/jbc.M313141200

7. Cho YS, Challa S, Moquin D, Genga R, Ray TD, Guildford M, Chan FK (2009) Phosphorylation-driven assembly of the RIP1-RIP3 complex regulates programmed necrosis and virus-induced inflammation. Cell 137(6):1112–1123. https://doi.org/10.1016/j.cell.2009.05.037
8. He S, Wang L, Miao L, Wang T, Du F, Zhao L, Wang X (2009) Receptor interacting protein kinase-3 determines cellular necrotic response to TNF-alpha. Cell 137(6):1100–1111. https://doi.org/10.1016/j.cell.2009.05.021
9. Zhao J, Jitkaew S, Cai Z, Choksi S, Li Q, Luo J, Liu ZG (2012) Mixed lineage kinase domain-like is a key receptor interacting protein 3 downstream component of TNF-induced necrosis. Proc Natl Acad Sci U S A 109(14):5322–5327. https://doi.org/10.1073/pnas.1200012109
10. Cai Z, Jitkaew S, Zhao J, Chiang HC, Choksi S, Liu J, Ward Y, Wu LG, Liu ZG (2014) Plasma membrane translocation of trimerized MLKL protein is required for TNF-induced necroptosis. Nat Cell Biol 16(1):55–65. https://doi.org/10.1038/ncb2883
11. Chen X, Li W, Ren J, Huang D, He WT, Song Y, Yang C, Li W, Zheng X, Chen P, Han J (2014) Translocation of mixed lineage kinase domain-like protein to plasma membrane leads to necrotic cell death. Cell Res 24(1):105–121. https://doi.org/10.1038/cr.2013.171
12. Wang H, Sun L, Su L, Rizo J, Liu L, Wang LF, Wang FS, Wang X (2014) Mixed lineage kinase domain-like protein MLKL causes necrotic membrane disruption upon phosphorylation by RIP3. Mol Cell 54(1):133–146. https://doi.org/10.1016/j.molcel.2014.03.003
13. Dondelinger Y, Declercq W, Montessuit S, Roelandt R, Goncalves A, Bruggeman I, Hulpiau P, Weber K, Sehon CA, Marquis RW, Bertin J, Gough PJ, Savvides S, Martinou JC, Bertrand MJ, Vandenabeele P (2014) MLKL compromises plasma membrane integrity by binding to phosphatidylinositol phosphates. Cell Rep 7(4):971–981. https://doi.org/10.1016/j.celrep.2014.04.026
14. Lu J, Bai L, Sun H, Nikolovska-Coleska Z, McEachern D, Qiu S, Miller RS, Yi H, Shangary S, Sun Y, Meagher JL, Stuckey JA, Wang S (2008) SM-164: a novel, bivalent Smac mimetic that induces apoptosis and tumor regression by concurrent removal of the blockade of cIAP-1/2 and XIAP. Cancer Res 68(22):9384–9393. https://doi.org/10.1158/0008-5472.CAN-08-2655

Chapter 9

Analysis of Cytokine- and Influenza A Virus-Driven RIPK3 Necrosome Formation

Roshan J. Thapa, Shoko Nogusa, and Siddharth Balachandran

Abstract

In multicellular organisms, regulated cell death plays a vital role in development, adult tissue homeostasis, and clearance of damaged or infected cells. Necroptosis is one such form of regulated cell death, characterized by its reliance on receptor-interacting protein kinase 3 (RIPK3). Once activated, RIPK3 nucleates a protein complex, termed the "necrosome," which includes the adaptors RIPK1 and FADD, and the effector protein MLKL. From the necrosome, RIPK3 phosphorylates MLKL to drive necroptosis, and can also induce RIPK1/FADD-mediated apoptosis, via caspase-8. Assembly of the necrosome thus serves as a useful readout of RIPK3 activation. In this chapter, we describe molecular methods for examining necrosome activation in response to the cytokines TNF-α, IFN-β, and IFN-γ, and upon infection with influenza A virus (IAV).

Key words Necrosome, Necroptosis, Necrosis, RIPK1, RIPK3, MLKL, FADD

1 Introduction

The kinase RIPK3 has emerged as a central mediator of a form of regulated necrosis termed necroptosis. Necroptosis can be triggered by multiple innate-immune pathways, including those activated by the cytokines tumor necrosis factor-α (TNF-α) [1, 2], type I (predominantly α/β) and type II (γ) interferons (IFNs) [3, 4], and the pathogen sensors DAI (ZBP1/DLM-1) [5, 6], toll-like receptor 3 (TLR3) [7, 8], and TLR4 [7, 8]. RIPK3 contains a RIP homology interaction motif (RHIM) [9]. Typically, the pathways that stimulate RIPK3 employ an upstream RHIM-containing protein that, upon activation, associates with RIPK3 via homotypic RHIM-RHIM-based interactions to nucleate this kinase into molecular signaling complex called the "necrosome" [9, 10]. For example, the RHIM-containing adaptors/sensors RIPK1, TRIF, and DAI bind and activate RIPK3 downstream of TNF-α, TLRs 3 and 4, and certain DNA/RNA virus infections, respectively [7, 11]. In most cases, a necrosome is formed that contains, at a minimum, RIPK3, RIPK1, FADD, and MLKL. From within the

Adrian T. Ting (ed.), *Programmed Necrosis: Methods and Protocols*, Methods in Molecular Biology, vol. 1857, https://doi.org/10.1007/978-1-4939-8754-2_9,

necrosome, RIPK3 phosphorylates MLKL, which then oligomerizes, translocates to cellular membranes, and disrupts these membranes to trigger necrotic death [12].

Notably, necroptosis is greatly potentiated by the inhibition of caspase activity, in particular, that of caspase-8, and activation of necroptosis downstream of certain stimuli (e.g., TNF-α) often requires inhibition of caspases [13]. But other stimuli, such as influenza A virus infection, activate necroptosis without the need for concurrent caspase blockade [14]. In these cases, RIPK3 can deploy not only MLKL (for the activation of necroptosis) but also caspase-8 for the induction of apoptosis [14]. The necrosome is thus a signaling platform for the activation of RIPK3-driven cell death pathways, and evidence of its formation serves as an excellent readout for the activation of this kinase.

In this chapter, we describe methods to detect necrosome formation upon stimulation with the cytokines TNF-α, type I IFNs, and type II IFN, and following infection of cells with influenza A virus. This protocol has been optimized for lysates from primary, early passage murine embryo fibroblasts, but works effectively on other RIPK3-competent murine cell types as well.

2 Materials

2.1 Culturing Mouse Embryonic Fibroblasts (MEFs)

1. Early passage ($P < 5$) primary wild-type (e.g., C57Bl/6) MEFs (*see* **Note 1**).
2. Complete DMEM: Dulbecco's Modified Eagle Medium supplemented with 15% heat-inactivated fetal bovine serum (FBS), 1% L-glutamine, 1% sodium pyruvate, 1% penicillin–streptomycin, and 1% Fungizone.
3. 60 mm cell culture dishes (Costar).

2.2 Cytokine Treatment or Influenza A Virus (IAV) Infection

1. Murine TNF-α (R&D Systems).
2. Murine IFN-γ (R&D Systems).
3. Murine IFN-β (Pestka Biomedical Laboratories).
4. Cycloheximide (Sigma).
5. zVAD.fmk (Bachem).
6. IAV strain A/Puerto Rico/8/1934 (PR8) (*see* **Note 2**).

2.3 Coimmunoprecipitation and Immunoblotting

1. Phosphate-buffered saline (PBS), 1×.
2. TL buffer: 1% Triton X-100, 150 mM Sodium chloride, 20 mM HEPES (pH 7.3), 5 mM EDTA, 5 mM Sodium fluoride, 0.2 mM Sodium ortho-vanadate, complete Mini® Protease inhibitor (one tablet per 10 mL, to be added right before use).
3. Cell scrapers.
4. Probe sonicator.

5. Protein A/G agarose beads (Thermo Scientific).
6. Micro-centrifuge tube rotator.
7. Anti-RIPK3 antibody (rabbit polyclonal; ProSci) (*see* **Note 3**).
8. SDS-PAGE loading dye: 0.5 M Tris (pH 6.8), 8% (w/v) SDS, 40% (v/v) glycerol, 0.1% (w/v) bromophenol blue, 1% (v/v) 2-mercaptoethanol.
9. Dry heating block.
10. SDS-PAGE gels (*see* **Note 4**).
11. Immunoblot equipment.
12. Nitrocellulose membrane (Millipore).
13. Nonfat dry milk.
14. Anti-RIPK1 antibody (mouse monoclonal; clone 38/RIP; BD Biosciences).
15. Anti-FADD antibody (mouse monoclonal; clone 1F7; EMD Millipore).
16. Anti-MLKL antibody (rat monoclonal; clone 3H1; EMD Millipore).
17. Species-specific secondary antibody conjugated to horseradish peroxidase (HRP) (Jackson ImmunoResearch).
18. Enhanced chemiluminescence (ECL) substrate (Thermo Scientific).

3 Methods

3.1 Cytokine Treatment (TNF-α, IFN-γ, IFN-β)

1. Plate 0.75×10^6 primary MEFs per 60 mm dish in 5 mL complete DMEM, using one dish per experimental condition. Gently rock the dishes back and forth to evenly spread cells, and incubate overnight in humidified incubator maintained at 37 °C/5% CO_2.
2. The next day, when MEFs are 80–90% confluent, replace medium with 5 mL fresh, prewarmed complete DMEM. Pretreat cells with the pancaspase inhibitor zVAD. fmk (hereafter called zVAD; 50 μM) for 1 h. When evaluating necrosome formation in response to TNF-α, also include cycloheximide (250 ng/mL) with zVAD (*see* **Note 5**).
3. Predilute 250 ng TNF-α in 100 μL warm serum-free DMEM and add to cells after they have been preincubated with zVAD and cycloheximide for 1 h. This will result in a final concentration of 50 ng/mL TNF-α. Similarly, predilute IFN-γ (50 ng) or IFN-β (125 ng) in 100 μL serum-free DMEM and add to cells that have been preincubated with zVAD, for a final concentration of 10 ng/mL IFN-γ and 25 ng/mL IFN-β, respectively. Incubate cells for 4–6 h before proceeding with coimmunoprecipitation.

3.2 Influenza A Virus Infection

1. Plate 0.75 × 10^6 primary MEFs per 60 mm dish in 5 mL complete DMEM, using one dish per experimental condition. Gently rock the dishes back and forth to evenly spread cells, and incubate overnight in humidified incubator maintained at 37 °C/5% CO_2.
2. The next day, when cells are 80–90% confluent, replace medium with prewarmed serum-free DMEM containing virus inoculum at the desired multiplicity of infection (MOI) in a volume that is sufficient to cover the monolayer cells. We typically prepare our virus inoculum in 1 mL serum-free medium per 60 mm dish, and infect cells at MOIs of between 2 and 5 for robust necrosome formation. Incubate cells with virus inoculum for 1 h in a humidified incubator maintained at 37 °C/5% CO_2 with frequent rocking (every 5–10 min) to keep the plates from drying out (*see* **Note 6**).
3. Aspirate the virus inoculum and wash the cells once gently with PBS. Add 5 mL complete growth medium to the cells and incubate them for 12–15 h (*see* **Note 7**).

3.3 Immunoprecipitation of RIPK3 and Detection of Necrosome Components

1. To immunoprecipitate RIPK3 from MEFs, gently wash the cells twice with PBS, and after thorough aspiration of PBS, lyse the cells by adding in 500 μL ice-cold TL buffer directly onto the plated cells and incubating the dish on ice for 5 min.
2. Scrape cells off the dish using a plastic cell scraper and collect lysates into a chilled microcentrifuge tube.
3. Sonicate the lysates with a 2-s-long pulse from a probe sonicator, and place the sample on ice for a minute. Repeat this step four times.
4. Clarify lysates by centrifuging at 10,000 × *g* for 5 min at 4 °C. Transfer the cleared lysates into a new microcentrifuge tube kept on ice, and discard the pellet.
5. Wash protein A/G agarose beads twice with TL buffer, and resuspend as a 50% w/v slurry in TL buffer. Each sample will require 60 μL slurry (30 μL for preclearing and 30 μL for immunoprecipitation).
6. To preclear the lysates, add 30 μL of protein A/G agarose slurry into each sample and incubate for an hour with rotation at 4 °C (*see* **Note 8**).
7. Centrifuge samples at 2500 × *g* for 5 min at 4 °C, carefully transfer supernatant into clean tube, and discard beads used for preclearing. At this step, save 25 μL (5% of total volume) lysate as input control.
8. Add 2 μg anti-RIPK3 polyclonal antibody to each sample and incubate overnight at 4 °C with rotation (*see* **Note 9**).
9. 12–16 h later, add 30 μL of twice-washed protein A/G agarose slurry into each tube and incubate samples for 2–3 h with rotation at 4 °C.

10. Wash beads three times with chilled TL buffer. For each wash, mix the beads with 1 mL TL lysis buffer, gently shake to disperse beads throughout the buffer, and centrifuge at 400 × *g* for 5 min at 4 °C. Following each wash, discard the supernatant by careful aspiration. After the final wash, remove as much TL buffer as possible from the protein A/G agarose beads.
11. Add 30 μL of SDS-PAGE loading buffer to each sample, denature by boiling for 5 min at 95 °C in a dry heating block, and centrifuge the samples at 10,000 × *g* for 5 min. Carefully collect the supernatant and discard the beads. Samples are now ready for immunoblot analysis and detection of necrosome components.
12. Separate samples by SDS-PAGE. Electrophorese the samples at 120 V and transfer onto nitrocellulose membrane as described previously [15] (*see* **Note 4**).
13. After soaking the transferred membrane in blocking buffer (0.1% [v/v] Tween20, 5% [w/v] nonfat dry milk in PBS) for 45 min, incubate the blot with RIPK1, FADD, and/or MLKL-specific monoclonal antibodies in blocking buffer for overnight at 4 °C with gentle rocking.
14. To detect primary antibodies, incubate the blot with species-specific secondary antibody–horseradish peroxidase (HRP) conjugates in blocking buffer for 2–3 h at room temperature followed by chemiluminescence (*see* Fig. 1).

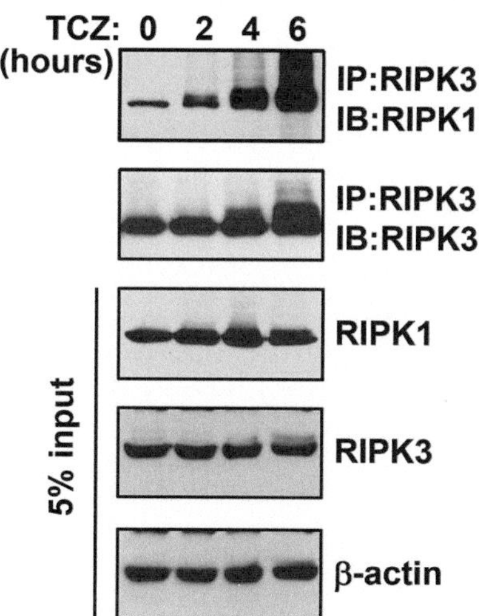

Fig. 1 Anti-RIPK3 immunoprecipitates from wild-type MEFs treated with TCZ [TNF-α (50 ng/mL) + cycloheximide (250 ng/mL) + z-VAD (50 μM)] for up to 6 h were examined for association between RIPK1 and RIPK3, as evidence of necrosome formation. In parallel, immunoblot analysis of input whole cell extract was performed to demonstrate equal amounts of protein in each lane. β-actin was used as a loading control

4 Notes

1. Primary murine embryo fibroblasts (MEFs) are prepared from day 13.5 embryos as previously described [16]. Check to ensure expression of key effector proteins (RIPK3, RIPK1, FADD, and MLKL) before conducting experiments. Many cell types, particularly with passaging or upon immortalization tend to lose expression of RIPK3 and/or MLKL.
2. A/Puerto Rico/8/34 (H1N1) is generated by reverse genetics as previously described [17]. Briefly, seed virus is injected into 10-day-old embryonated hen's eggs, which are incubated at 37 °C for 48 h and chilled at 4 °C for a further 24 h. Allantoic fluid containing virus is then harvested from these eggs and characterized for titer on Madin-Darby Canine Kidney (MDCK) cells.
3. Anti-RIPK3 goat polyclonal antibody from Santa Cruz Biotechnology can serve as an alternative to the anti-RIPK3 rabbit polyclonal antibody from ProsSci. If using this antibody, use 5 μg antiserum for immunoprecipitation.
4. We routinely use 8% gels for detection of RIPK1, while 10–12% gels are used for detection of MLKL and FADD.
5. For formation of the necrosome after cytokine treatment, addition of zVAD is required.
6. A low dose of zVAD (5–10 μM) potentiates IAV-driven necrosome formation, but its addition is not essential.
7. We typically harvest IAV-infected cells when ~50% of the monolayer displays some early cytopathic effect (CPE). A classic sign of early CPE is cell-rounding (but not detachment), accompanied by increased refractility. At MOI = 2, we observe CPE in MEFs by 12–15 h of infection. If CPE is not detected, we encourage delaying harvesting of cells until CPE becomes evident, up to ~24 h post infection.
8. Lysates are generally precleared to reduce nonspecific binding and minimize background.
9. 2 μg rabbit IgG control antibody may be used as a negative control to rule out nonspecific association of necrosome components with protein A/G agarose beads.

References

1. Moriwaki K, Chan FK (2013) RIP3: a molecular switch for necrosis and inflammation. Genes Dev 27(15):1640–1649. https://doi.org/10.1101/gad.223321.113
2. Vandenabeele P, Galluzzi L, Vanden Berghe T, Kroemer G (2010) Molecular mechanisms of necroptosis: an ordered cellular explosion. Nat Rev Mol Cell Biol 11(10):700–714. https://doi.org/10.1038/nrm2970
3. McComb S, Cessford E, Alturki NA, Joseph J, Shutinoski B, Startek JB, Gamero AM, Mossman KL, Sad S (2014) Type-I interferon signaling through ISGF3 complex is required for sustained Rip3 activation and necroptosis in macrophages. Proc Natl Acad Sci U S A 111(31):E3206–E3213. https://doi.org/10.1073/pnas.1407068111
4. Thapa RJ, Nogusa S, Chen P, Maki JL, Lerro A, Andrake M, Rall GF, Degterev A,

Balachandran S (2013) Interferon-induced RIP1/RIP3-mediated necrosis requires PKR and is licensed by FADD and caspases. Proc Natl Acad Sci U S A 110(33):E3109–E3118. https://doi.org/10.1073/pnas.1301218110

5. Mocarski ES, Kaiser WJ, Livingston-Rosanoff D, Upton JW, Daley-Bauer LP (2014) True grit: programmed necrosis in antiviral host defense, inflammation, and immunogenicity. J Immunol 192(5):2019–2026. https://doi.org/10.4049/jimmunol.1302426
6. Sridharan H, Upton JW (2014) Programmed necrosis in microbial pathogenesis. Trends Microbiol 22(4):199–207. https://doi.org/10.1016/j.tim.2014.01.005
7. Pasparakis M, Vandenabeele P (2015) Necroptosis and its role in inflammation. Nature 517(7534):311–320. https://doi.org/10.1038/nature14191
8. He S, Liang Y, Shao F, Wang X (2011) Toll-like receptors activate programmed necrosis in macrophages through a receptor-interacting kinase-3-mediated pathway. Proc Natl Acad Sci U S A 108(50):20054–20059. https://doi.org/10.1073/pnas.1116302108
9. Sun X, Yin J, Starovasnik MA, Fairbrother WJ, Dixit VM (2002) Identification of a novel homotypic interaction motif required for the phosphorylation of receptor-interacting protein (RIP) by RIP3. J Biol Chem 277(11):9505–9511. https://doi.org/10.1074/jbc.M109488200
10. Cho YS, Challa S, Moquin D, Genga R, Ray TD, Guildford M, Chan FK (2009) Phosphorylation-driven assembly of the RIP1-RIP3 complex regulates programmed necrosis and virus-induced inflammation. Cell 137(6):1112–1123. https://doi.org/10.1016/j.cell.2009.05.037
11. Wallach D, Kang TB, Dillon CP, Green DR (2016) Programmed necrosis in inflammation: toward identification of the effector molecules. Science 352(6281):aaf2154. https://doi.org/10.1126/science.aaf2154
12. Zhang J, Yang Y, He W, Sun L (2016) Necrosome core machinery: MLKL. Cell Mol Life Sci 73(11–12):2153–2163. https://doi.org/10.1007/s00018-016-2190-5
13. Sun L, Wang X (2014) A new kind of cell suicide: mechanisms and functions of programmed necrosis. Trends Biochem Sci 39(12):587–593. https://doi.org/10.1016/j.tibs.2014.10.003
14. Nogusa S, Thapa RJ, Dillon CP, Liedmann S, Oguin TH 3rd, Ingram JP, Rodriguez DA, Kosoff R, Sharma S, Sturm O, Verbist K, Gough PJ, Bertin J, Hartmann BM, Sealfon SC, Kaiser WJ, Mocarski ES, Lopez CB, Thomas PG, Oberst A, Green DR, Balachandran S (2016) RIPK3 activates parallel pathways of MLKL-driven necroptosis and FADD-mediated apoptosis to protect against influenza a virus. Cell Host Microbe 20(1):13–24. https://doi.org/10.1016/j.chom.2016.05.011
15. Chen P, Nogusa S, Thapa RJ, Shaller C, Simmons H, Peri S, Adams GP, Balachandran S (2013) Anti-CD70 immunocytokines for exploitation of interferon-gamma-induced RIP1-dependent necrosis in renal cell carcinoma. PLoS One 8(4):e61446. https://doi.org/10.1371/journal.pone.0061446
16. Conner DA (2001) Mouse embryo fibroblast (MEF) feeder cell preparation. Curr Protoc Mol Biol Chapter 23:Unit 23 22. doi:https://doi.org/10.1002/0471142727.mb2302s51
17. Hoffmann E, Neumann G, Kawaoka Y, Hobom G, Webster RG (2000) A DNA transfection system for generation of influenza a virus from eight plasmids. Proc Natl Acad Sci U S A 97(11):6108–6113. https://doi.org/10.1073/pnas.100133697

Chapter 10

Detection of RIPK1 in the FADD-Containing Death Inducing Signaling Complex (DISC) During Necroptosis

Rosalind L. Ang and Adrian T. Ting

Abstract

FAS-associated protein with death domain (FADD) is a signaling molecule required by members of the TNF receptor superfamily (TNFRSF) such as FAS and TNFR1 to induce apoptosis. FADD is a small adapter molecule that functions as a scaffold to recruit procaspase-8 and other regulators. The FADD-containing signaling complex that initiates the apoptotic cascade has been termed the death inducing signaling complex (DISC). In the absence of FADD, death receptors cannot induce apoptosis and in appropriate cell types, these death receptors then induce necroptosis. Necroptosis can also be induced by death receptors in FADD-sufficient cells when caspase-8 is inhibited, usually accomplished by the addition of caspase inhibitors. Under such necroptotic conditions, the immunoprecipitation of FADD to isolate the DISC can be utilized to examine components of this complex. Here, we describe the immunoprecipitation of FADD and subsequent western-blotting to identify RIPK1 in this complex during necroptosis.

Key words Death inducing signaling complex (DISC), FAS-associated protein with death domain (FADD), Receptor-interacting protein kinase 1 (RIPK1), Coimmunoprecipitation, Western blotting

1 Introduction

FADD, also known as MORT1, plays diverse roles in proliferation, inflammation, innate immunity and autophagy (reviewed in [1]). It has been well studied in the signaling of death receptors of the TNFRSF where it serves as an adaptor molecule. FADD has a bipartite structure. The N-terminal domain encodes a death effector domain (DED), which interacts homotypically with DED present in apical caspases such as procaspse-8 and procaspase-10. The C-terminal end of FADD contains a death domain (DD), which interacts with the DD present in death receptors such as FAS or TNFR signaling molecules such as TRADD. FADD was first described to initiate apoptosis by forming a bridge between FAS and procaspase-8 following receptor activation. procaspase-8 undergoes self-cleavage when brought into close proximity to form a tetrameric p18/p10 active enzyme to initiate cell death. This cytoplasmic FADD-caspase-8 complex was

Adrian T. Ting (ed.), *Programmed Necrosis: Methods and Protocols*, Methods in Molecular Biology, vol. 1857, https://doi.org/10.1007/978-1-4939-8754-2_10,

termed the death inducing signaling complex (DISC) [2]. These molecular interactions are not specific to FAS signaling. In TNFR1 signaling, FADD is observed to interact with TRADD that are released from TNFR1 complex I to recruit procaspase-8 to initiate death and this complex is known as complex II [3].

The homozygous ablation of the *Fadd* gene in mice results in embryonic lethality at E10.5, evidenced by massive necrosis [4, 5]. This embryonic lethality was rescued by deleting either *Ripk1* [6] or *Ripk3* [7], which are essential for necroptosis induced by TNFR1. The *Fadd*$^{-/-}$*Ripk3*$^{-/-}$ double-deficient embryos are viable and develop normally [7]. The phenotype of the *Fadd*$^{-/-}$ mice mirrors that of *Casp8*$^{-/-}$ mice and these in vivo genetic ablation experiments revealed that FADD and caspase-8 are not only essential for inducing apoptotic death but are also involved in keeping necroptosis in check.

Cells that are deficient in either FADD or caspase-8 undergo necroptosis in response to TNF. However in FADD or caspase-8-sufficient cells, they can be induced to undergo necroptosis in the presence of a caspase inhibitor and one that is frequently used is zVAD-FMK. In some cell types such as mouse embryonic fibroblasts (MEF), bone marrow derived macrophages (BMDM) or the L929 cell line, stimulating with TNF in the presence of zVAD-FMK is sufficient to trigger necroptosis. In other cell types, it may be necessary to combine TNF, zVAD-FMK together with another agent such as SMAC mimetics or cycloheximide to induce necroptosis. RIPK1 is necessary for necroptosis induced by TNF and in the presence of zVAD-FMK, it forms a stable complex with FADD and caspase-8. This complex can be isolated by immunoprecipitation with anti-FADD and this is sometimes termed the “FADD necrosome.” Since necroptosis clearly occurs in the absence of FADD, the term “FADD necrosome” may be inappropriate and the term necrosome should best be used to describe the core RIPK1-RIPK3-MLKL necroptotic complex. In addition, this RIPK1-FADD-caspase-8 complex is also formed during apoptosis, which is sometimes referred to as the “ripoptosome.” Because this complex has the potential to initiate either apoptosis or necroptosis depending on the cell type and experimental conditions, we prefer to call this the DISC.

In this chapter, we detail the protocol for immunoprecipitating the DISC with FADD specific antibody followed by the detection of RIPK1 in this complex. This signaling event is a hallmark of cells undergoing RIPK1-dependent cell death. We describe here the protocol used to detect DISC in MEF and in BMDM.

2 Materials

Prepare all solutions with ultrapure water (prepared by purifying deionized water (18 MΩ-cm at 25 °C)) and molecular grade reagents.

2.1 Preparation of the Lysate

1. Lysis buffer: 20 mM Tris–HCl pH 7.5, 150 mM NaCl, 10% glycerol, 0.2% NP-40. Keep at 4 °C. Add the following fresh before use: 0.1 mM sodium orthovanadate, 5 mM β-glycerophosphate and protease inhibitors.
2. Phosphate-buffered Saline (PBS).
3. Mouse TNF (R & D System). Stock is prepared by reconstituting lyophilized TNF in sterile PBS with 0.1% bovine serum albumin at a concentration of 100 μg/mL.
4. Cycloheximide: 50 mg/mL in ethanol.
5. z-Val-Ala-Asp-(OMe)-Fluoromethyl Ketone (z-VAD-FMK): stock concentration at 50 mM in DMSO.

2.2 Reagents and Instrument for Immunoprecipitation

1. FADD antibody (M19, Santa Cruz Biotechnology).
2. Protein A/G beads (Santa Cruz Biotechnology).
3. 2× denaturing sample buffer: 65.8 mM Tris–HCl pH 6.8, 26.3% glycerol, 2.1% SDS, 0.01% bromophenol blue. Store at room temperature. Before use, add 50 μL β-mercaptoethanol to 950 μL 2× sample buffer.
4. Microfuge tube rotator placed in cold room.
5. 70 °C heat block

2.3 Electrophoresis and Immunoblotting

1. 10% SDS PAGE gel
2. TBST: 25 mM Tris base, 137 mM NaCl, 2.7 mM KCl, 0.02% Tween 20.
3. Nonfat dry milk powder.
4. 0.45 μm nitrocellulose membrane.
5. Lab rocker.

3 Methods

3.1 Cell Stimulation

1. Seed 2×10^6 MEF onto 10 cm tissue culture plate.
2. The next day, check that the cells are nicely spread out.
3. Then remove the media and replace it with 8 mL of media containing 20 μM zVAD and 1 μg/mL cycloheximide (*see* **Note 1**).
4. Incubate at 37 °C for 1 h in CO_2 incubator.
5. Add mTNF to a concentration of 100 ng/mL and incubate for 2 or 4 h at 37 °C in a CO_2 incubator. The control plate is left unstimulated.
6. Put the plate on ice at the end of incubation and aspirate the media.
7. Pipet 10 mL of cold PBS into the plate. Swirl it three times and aspirate the PBS (*see* **Note 2**).

3.2 Cell Lysis and Immunoprecipitation

1. Add 1 mL of chilled lysis buffer.
2. Use a cell scraper to gently break the cells and pipet the suspension into a chilled microfuge tube.
3. Place the microfuge tube on ice for 20 min.
4. Spin at 9400 × *g* in a microcentrifuge for 15 min.
5. Pipet and transfer the supernatant into a fresh chilled microfuge tube. Leave the pellet untouched.
6. Measure the total protein amount with colorimetric assay such as Pierce BCA protein assay.
7. Transfer equivalent amount of protein from each sample into chilled microfuge tubes and add 1 μg of FADD antibody to the lysate.
8. Place the microfuge tubes on a rotator to rotate overnight at 4 °C.

3.3 Preparation of Protein A/G Beads

1. The next day, pipet the required amount of A/G beads into a chilled microfuge tubes. 30 μL of A/G bead slurry is required for each sample. For example, if you are going to analyze four samples, you should prepare enough A/G beads for five samples, i.e., 150 μL of beads.
2. Add 1 mL chilled lysis buffer (prepared fresh) to A/G beads and centrifuge at 1500 × *g* for 30 s in a chilled microcentrifuge.
3. Remove the supernatant with a 1 mL pipette. Then repeat **step 2** for a total of four times. These washing steps are to remove the preservative and buffer used for long-term storage of the A/G beads.
4. At the last washing step, retain 100 μL of the lysis buffer. Microfuge briefly and remove the remaining 100 μL of supernatant with a 200 μL pipette.
5. Add appropriate amount of chilled lysis buffer and aliquot equal volume of A/G bead suspension into fresh chilled microfuge tubes. If a 200 μL pipette is used, cut the tip of the pipette tip before using it to aliquot the suspension.
6. Centrifuge the A/G bead suspension at 1500 × *g* for 30 s in a chilled microcentrifuge. Inspect that equal amounts of pellet are in the tubes.

3.4 Isolation and Washing of FADD Immune Complexes

1. Remove the microfuge tubes containing the lysates with the FADD antibody from the rotator and centrifuge at 9400 × *g* in a chilled microcentrifuge.
2. Transfer the supernatants into the washed A/G beads suspension and rotation for 3 h at 4 °C.
3. Pellet the A/G beads by spinning at 1500 × *g* for 30 s in a chilled microcentrifuge.

4. Discard the supernatant with 1 mL pipette. Aspirate out as much as possible. Then add 1 mL of chilled lysis buffer to resuspend the beads. Spin the microfuges tubes at 1500 × *g* for 30 s in a chilled microcentrifuge.
5. Repeat **step 4** for four more times. These washing steps are to remove unbound proteins and allow nonspecific proteins to diffuse from the interior of the beads. These washes should be done over at least 30 min total. Suggested time is 5 min per wash step.
6. At the final wash step, leave behind approximately 100 μL of the lysis buffer. Microcentrifuge briefly to collect the remaining buffer and remove as much of it as possible.
7. Prepare 2× sample buffer with beta-mercaptoethanol and pipet 10–20 μL into each tube.
8. Incubate the samples in the 70 °C heat block for 20 min to elute the proteins from the A/G beads.
9. Place the tubes at room temperature for at least 5 min to cool down.
10. Spin at 9400 × *g* in microcentrifuge for a minute. Collect and transfer the supernatant (containing the denatured proteins) to a fresh tube. Alternatively, transfer the bead and sample buffer mixture into a filter tube and spin at high speed in microcentrifuge to separate samples from the A/G beads. At this stage, samples may be stored at −80 °C for future analysis.

3.5 SDS-PAGE and Anti-RIPK1 Blotting

1. Run the denatured samples on 10% SDS-PAGE gel and transfer to nitrocellulose membrane for immunoblotting.
2. Following the transfer, place the membrane in a container and add generous amount of Tris-buffered saline with 0.2% Tween (TBST). Place the container on a platform shaker for 10 min at room temperature to wash the membrane. Discard the TBST.
3. Repeat the wash two more times.
4. Add sufficient amount of TBST with 5% milk to cover the membrane and rock for an hour at room temperature to block the membrane (*see* **Note 3**).
5. Add sufficient amount of TBST with 5% BSA containing diluted RIPK1 antibody and rock overnight at 4 °C (*see* **Note 4**).
6. Wash the membrane three times as in **steps 2** and **3**.
7. Add sufficient amount of TBST with 5% milk containing diluted secondary antibody conjugated to horseradish peroxidase or infrared fluorescence dye. Rock the membrane for an hour at room temperature.
8. Wash the membrane three times as in **steps 2** and **3**. The membrane is now ready for detection using the appropriate signal

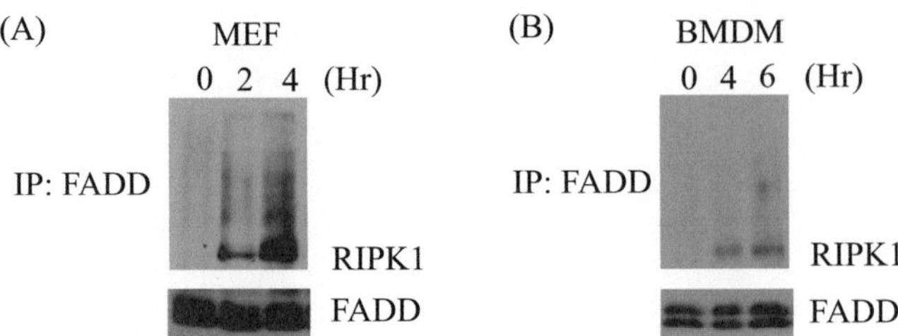

Fig. 1 (**a**) Wild-type MEF were stimulated with mTNF (100 ηg/mL) in the presence of cycloheximide (1 μg/mL) and zVAD (20 μM) for the time specified. (**b**) BMDM were stimulated with LPS (10 ηg/mL) in the presence of cycloheximide (1 μg/mL) and zVAD (20 μM) for the time specified. In both panels, lysates were then subjected to FADD immunoprecipitation to isolate the death-inducing signaling complex (DISC) followed by blotting for RIPK1

detection system. Figure 1 shows the presence of RIPK1 in the DISC isolated from MEF after 2 and 4 h of TNF stimulation (Panel A) and in bone marrow derived macrophages after 4 and 6 h of LPS stimulation (Panel B).

9. To reprobe, the membrane can be stripped with 6 M guanidine hydrochloride for 10 min and then washed three times with TBST for 10 min each and then block with TBST with 5% milk.

4 Notes

1. The cycloheximide sensitizes cells to TNF-induced death by preventing the synthesis of survival gene products. The pancaspase inhibitor zVAD-FMK serves two purposes. One, it functionally switches the death response from apoptosis to necroptosis. Two, it stabilizes the DISC complex because it prevents caspase-8 present in the FADD complex from cleaving RIPK1. This allows RIPK1 to be more readily detected.
2. The procedure can be stopped at this point. Seal the plates with parafilm and stored at −80 °C for future processing.
3. Overnight incubation at 4 °C can be performed. If short of time, a minimum of 30 min incubation at room temperature is sufficient.
4. If short of time, this step can be performed at room temperature with 1–2 h of rocking.

Acknowledgments

This work was supported by grants AI052417 and DK072201 from the NIH, and a Senior Research Award from the Crohn's and Colitis Foundation of America (CCFA).

References

1. Tourneur L, Chiocchia G (2010) FADD: a regulator of life and death. Trends Immunol 31:260–269
2. Kischkel FC, Hellbardt S, Behrmann I, Germer M, Pawlita M, Krammer PH, Peter ME (1995) Cytotoxicity-dependent APO-1 (Fas/CD95)-associated proteins form a death-inducing signaling complex (DISC) with the receptor. EMBO J 14:5579–5588
3. Hsu H, Shu HB, Pan MG, Goeddel DV (1996) TRADD-TRAF2 and TRADD-FADD interactions define two distinct TNF receptor 1 signal transduction pathways. Cell 84:299–308
4. Yeh WC, de la Pompa JL, McCurrach ME, Shu HB, Elia AJ, Shahinian A, Ng M, Wakeham A, Khoo W, Mitchell K, El-Deiry WS, Lowe SW, Goeddel DV, Mak TW (1998) FADD: essential for embryo development and signaling from some, but not all, inducers of apoptosis. Science 279:1954–1958
5. Zhang J, Cado D, Chen A, Kabra NH, Winoto A (1998) Fas-mediated apoptosis and activation-induced T-cell proliferation are defective in mice lacking FADD/Mort1. Nature 392:296–300
6. Zhang H, Zhou X, McQuade T, Li J, Chan FK, Zhang J (2011) Functional complementation between FADD and RIP1 in embryos and lymphocytes. Nature 471:373–376
7. Dillon CP, Oberst A, Weinlich R, Janke LJ, Kang TB, Ben-Moshe T, Mak TW, Wallach D, Green DR (2012) Survival function of the FADD-CASPASE-8-cFLIP(L) complex. Cell Rep 1:401–407

Chapter 11

Use of RIP1 Kinase Small-Molecule Inhibitors in Studying Necroptosis

Allison M. Beal, John Bertin, and Michael A. Reilly

Abstract

RIP1 kinase plays a key role in regulating signaling pathways downstream of a number of innate immune receptors such as TNFRI and TLRs. The discovery of Necrostatin-1 (Nec-1) as a small-molecule inhibitor of RIP1 kinase has been very instrumental in defining the necroptotic and other signalling pathways regulated by RIP1, but certain characteristics of Nec-1 limits its utility in experimental systems. Next generation RIP1 kinase inhibitors have been identified and the use of these tool inhibitors along with Nec-1 has revealed that RIP1 is emerging as a key driver of inflammation and tissue injury in the pathogenesis of various diseases. Further studying the role of RIP1 to carefully unravel the complex biology requires the selection of the correct tool small-molecule inhibitors. In addition, it is important to consider the proper application of current tool inhibitors and understand the current limitiations. Here we will discuss key parameters that need to be considered when selecting and applying tool inhibitors to novel biological assays and systems. General protocols to explore the in vitro and in vivo potency, cellular selectivity, and pharmacokinetic properties of current small-molecule inhibitors of RIP1 kinase are provided.

Key words RIP1 kinase, Necroptosis, GSK'963, GSK'481, Nec-1

1 Introduction

RIP1 kinase has emerged as a key node in regulation of various cellular processes underlying the pathogenesis of numerous inflammatory diseases. Targeting RIP1 kinase to further understand RIP1 biology comes with the challenge of having to discriminate between the independent scaffolding and kinase functions. This limits the utility of traditional genetic knockout or knockdown methods in exploring RIP1 kinase activity [1–3]. RIP1 kinase-dead knockin mice have been described but recent evidence suggests that RIP1 kinase inhibitors may do more than block kinase activity [4]. Careful interrogation of RIP1 function in in vitro and in vivo models requires the appropriate tools making small-molecule inhibitors of RIP1 an ideal choice. Several inhibitors of RIP1 have been described in the literature. These include two representative

Adrian T. Ting (ed.), *Programmed Necrosis: Methods and Protocols*, Methods in Molecular Biology, vol. 1857, https://doi.org/10.1007/978-1-4939-8754-2_11,

molecules from two different chemical series from our labs which are both potent and selective inhibitors (GSK'963 [5] and GSK'481 [6]). It is imperative that inhibitors with favorable key parameters are considered and caution should be used to ensure the selection of the right inhibitors at the appropriate dose for use in novel models [4]. Ideal inhibitors should have the desirable properties of kinase selectivity, limited off-target activity, good potency, and good pharmacokinetic characteristics. Current tool inhibitors are good options for in vitro assays and acute (1–2 day models) in vivo models but are not ideal for use in chronic models. When considering the use of RIP1 kinase inhibitors, it is essential to select the right tools for careful and comprehensive interrogation of RIP1 biology. Here we will discuss the key considerations and methods for the characterization of inhibitors by exploring the potency, cellular selectivity and pharmacokinetic properties of known small-molecule inhibitors of RIP1 kinase for careful interrogation of RIP1 biology in new experimental systems.

2 Materials

Table 1 lists some of the properties of the currently available RIP1 kinase inhibitors.

2.1 In Vitro Generalized Cell-Based Necroptosis Assay

1. RIP1 kinase inhibitor stock solution: made in DMSO usually at 1–10 mM concentration.
2. QVD-Orph from Calbiochem: 50 μM final concentration for use in assay.
3. Z-Val-Ala-DL-Asp-fluoromethylketone (z-VAD-fmk) (Bachem or Promega).
4. mTNFa (R&D Systems): Final concentration for assay 50 ng/mL.
5. 96-well flat bottom plates compatible with luminometer.
6. Multichannel pipette.
7. CellTiter-Glo (CTG) Luminescent Cell Viability assay (Promega).
8. Plate shaker.
9. Luminometer.

2.1.1 Mouse Fibrosarcoma L929 Cell Culture

1. L929 cells ATCC# CCL-1.
2. Culture media: RPMI-1640 with phenol red without Hepes with 2.1 mM L-glutamine, 10% FBS, and P/S/A (penicillin, streptomycin, amphotericin B (antimycotic)) (*see* **Note 1**).
3. Trypsin (0.25% with phenol red) for normal passaging of cells. Versene (Thermo Fisher) is recommended for use on study days for a more gentle dissociation of cells.
4. Cell culture flasks.

Table 1
Currently available tool Inhibitors (*see* Note 12)

Allosteric Structural Templates				"Hinge-binding Templates"
Nec-1/Nec-1s [5, 11–14]	**GSK'481 [6]**	**GSK'963 [5]**	**RIPA-56 [15]**	**i.e., Ponatinib [16]**
Modest potency (micro molar potency)	Very potent in primates; > 100-fold less potent in nonprimates	Very potent (nano molar potency)	Very potent	Lack of kinase selectivity
Excellent selectivity	Excellent selectivity (10,000× over other kinases)	Excellent selectivity (10,000× over other kinases)	Excellent selectivity	Limited utility because of significant off target activities
Off-target activity/ improved off target activity Poor in vivo pharmacokinetic properties, precluding use in long term rodent models/limited in vivo pharmacokinetic properties, restricting use in long term rodent models	Very limited off target activity Acceptable PK profile in rodents but potency restricts use in long-term rodent efficacy models	Very limited off-target activity Limited in vivo pharmacokinetic properties, restricting use in long term rodent efficacy models	Very limited off-target activity Improved in vivo pharmacokinetic properties but still not optimal for use in long-term rodent efficacy models	
Nec-1s preferred over Nec-1: Useful for in vitro assays	Optimal for use in human cellular systems	Useful for in vitro assays and acute in vivo models; inactive enantiomer available for use as control	Useful for in vitro and acute in vivo models	Not recommended for in vitro or in vivo assays

2.1.2 Bone Marrow-Derived Macrophage Generation

1. Dissection tools: sharp scissors and forceps.
2. Ethanol.
3. HBSS, Hank's Balanced Salt Solution.
4. BMDM Complete Media: DMEM media, with high glucose (4500 mg/L), 4 mM L-glutamine and 25 mM Hepes + 10% heat-inactivated (HI) FBS + P/S/A.
5. M-CSF (R&D Systems): Reconstitute at 0.1 μg/mL for a stock solution that is aliquoted into 10 μL aliquots for addition

to 100 mL media. Final working concentration in BMDM media = 10 ng/mL.

6. Disposable 10 mL syringe and 23G needle.
7. Cell culture dishes.

2.2 Cell-Based Selectivity Assays

2.2.1 Apoptosis Assay

1. Cyclohexamide (CHX) (sigma): final concentration for assay = 12 μg/mL.
2. mTNFa (R&D System): Final concentration for assay 50 ng/mL.
3. Caspase-Glo 3/7 assay (Promega).
4. White-walled 96-well luminometer plates.
5. Multichannel pipette.

2.2.2 NF-κB Activation

1. mTNFa (R&D System): Final concentration for assay 50 ng/mL.
2. 10× RIPA buffer (Millipore). Make buffer 1× by adding milliQ water and add protease inhibitor tablet and phosphate inhibitors. Chill on ice prior to lysis of cells.
3. BCA assay for determination of protein concentration.
4. Sample loading buffer.
5. 4–12% SDS-PAGE gel.
6. MOPS buffer or appropriate running buffer for system of choice.
7. Nitrocellulose membrane.
8. Tris buffered saline (TBS) and TBS with Tween (TBST). TBST is made by adding Tween 20 at final concentration of 0.1%.
9. Antibodies for detection of NF-κB activation: IκB (Cell Signaling), phospho-IκB (Cell Signaling), and tubulin as a loading control.

2.3 Pharmacokinetic (PK) and Pharmacodynamic (PD) Assays to Determine Dose for In Vivo Disease Models

2.3.1 PK Assay

1. For dosing, inhibitors are prepared with the following stock solutions: DMSO, 40% Cavitron (2-hydroxypropyl)-β-cyclodextrin) in water or PBS, and sterile endotoxin-free water or 0.9% sterile saline (*see* **Note 2** about Cavitron).
2. Oral dosing: 10 mL/kg dose volume. 5% final DMSO concentration, 6% final Cavitron concentration. Sonicate for 5 min after the addition of DMSO and Cavitron. Add sterile water to final concentration.
3. IV or IP dosing: 5 mL/kg or 10 mL/kg dose volume. 5% final DMSO concentration, 6% final Cavitron concentration. Sonicate for 5 min after the addition of DMSO and Cavitron. Add 0.9% sterile saline to final concentration.

4. Dosing needles: Gavage needles for oral dosing, 27G needles for IV or IP administration.
5. 1 mL syringe.
6. Bullet tubes.
7. Water: HyClone molecular biology grade (GE Healthcare).
8. 25 μL capillary tubes.
9. Sharp scissors or razor blade for tail tip amputation.
10. Gauze pads.

2.3.2 PD Assay

1. Water: HyClone molecular biology grade (GE Healthcare).
2. PBS with 1 mg/mL bovine serum albumin (BSA).
3. Sterile 0.9% saline.
4. zVAD-fmk (Bachem) Formulation: Reconstitute one 25 mg vial of zVAD with 0.25 mL of DMSO to make a 100 mg/mL stock. Add appropriate volume of 100 mg/mL stock into a sterile 1.5 mL eppendorf tube containing appropriate volume of PBS containing 1 mg/mL BSA so that final concentration of zVAD is 4 mg/mL. Vortex thoroughly. The dose volume is adjusted according to body weight so that each mouse gets 16.67 mg/kg of zVAD at a 4.17 mL/kg dose volume. For example, each 24 g mouse will receive 0.1 mL for a final dose of 0.4 mg of zVAD-fmk.
5. zVad-fmk + TNFα (Cell Sciences Inc., Specific Activity>1×10^7 IU/mg) Formulation: Reconstitute each 1 mg vial of TNFα in 1 mL of sterile endotoxin free water containing 1 mg/mL BSA. The unused portion of TNFα can be frozen at −80 °C for 1 freeze thaw cycle. Add appropriate volume of 1 mg/mL TNFα to an appropriate volume of 100 mg/mL stock of zVAD-fmk. Bring up to final volume with sterile PBS containing 1 mg/mL BSA for a final concentration of 300 μg/mL of TNFα and 4 mg/mL of zVAD. Dose volume will be adjusted according to body weight such that each mouse gets 1.25 and 16.67 mg/kg of TNF and zVAD, respectively, at a 4.17 mL/kg dose volume. For example, each 24 g mouse will receive 0.1 mL for a final dose of 0.4 mg of zVAD-fmk and 30 μg TNFα.
6. Water bath.
7. 1 mL syringes with 27G needles.
8. Rectal thermometer.
9. Mouse restrainer for IV dosing.
10. Gauze pads.

3 Methods

3.1 Generalized Cell-Based Necroptosis Assay (In Vitro Potency Measure)

Generalized cell-based necroptosis assay can be used to define the half-maximal inhibitory concentration (IC50) of known small-molecule inhibitors of RIP1 in novel cellular systems. When exploring the role of RIP1 in new cell types and experimental systems where the biology of RIP1 is not well defined it is recommended that well-defined cellular systems such as the L929 mouse fibrosarcoma cell line or primary BMDM cells be used as controls for inducing necroptosis before applying the assay in new cell types (*see* **Note 3**). For exploration of necroptosis in human cellular systems, necroptosis induction in HT29 cells using TNFα/Smac mimetic/z-VAD-FMK (TSZ) can be used and has been well described elsewhere [1, 7–9].

Prepare cells for use in cell-based necroptosis assay. Cell lines or primary cells can be used. Conditions for culturing mouse fibrosarcoma L929 cells and differentiating BMDMs from mouse bone marrow are provided as an example.

3.1.1 L929 Cell Culture

1. Mouse fibrosarcoma L929 cells are cultured in complete RPMI media at 37 °C, 5% CO_2 per instructions from ATCC.
2. Cells are split the day of the assay to prepare for the necroptosis assay. L929 cells are adherent cells and thus need to be dissociated from the flask by using versene.
3. To split the cells remove the media and wash the flask gently with 10 mL PBS. Add 2–3 mL versene to the adherent cells and incubate for about 5 min or until all the cells have detached.
4. Once the cells have detached add 5–10 mL complete RPMI media and remove the cells from the flask.
5. Spin cells to remove versene, bring the cells up to 10 mL and then count the cells.
6. Plate cells appropriately for the necroptosis assay. For a typical necroptosis assay, 50,000 cells/well are plated in a 96-well flat bottom plate in preparation for the assay.

3.1.2 Preparation of Bone Marrow-Derived Macrophages (BMDM)

Except for the collection of the legs from the mice, all of the steps are performed in a sterile hood.

1. BMDM are prepared from bone marrow isolated from the hind limbs of mice. It is recommended to use the limbs for the same strain of mice as the mouse strain used for in vivo assays.
2. On day 0, bone marrow is isolated from the hind legs of mice. Hind legs are collected from euthanized mice after spraying with 70% ethanol. The skin is cut at the hind leg and carefully pulled back to expose the muscle. Cut the hind leg at the pelvic joint and place the legs in a 50 mL conical tube filled with DMEM.

3. Remove excess muscle from legs without breaking the bones and place the bones in 70% ethanol for 1–2 min.
4. Carefully isolate the femur (and tibia) by cutting at the knee joint being careful not to remove the epiphyses (proximal and distal ends of femur). Place in sterile PBS to wash off the ethanol until ready to flush.
5. To flush, remove epiphyses using a razor or scissors. Once the epiphyses are removed use a syringe and needle to flush the cells with DMEM out of the bone and into a sterile 50 mL conical tube. The bone will appear white when all of the marrow has been flushed. Keep the cells on ice after flushing.
6. Centrifuge the cells at 350 × g for 10 min at 4 °C.
7. Remove the supernatant and resuspend cells in 10 mL complete DMEM with 10 ng/mL M-CSF.
8. Count the cells to determine cell number. The average yield is ~10 × 10^7 cells per mouse. Cells are plated at 1 × 10^7 cells in 10 mL in a 10 cm tissue culture dish.
9. Incubate the cells at 37 °C and 2% CO_2.
10. The next day add 10 mL fresh complete DMEM media + M-CSF to each dish.
11. Every 2–3 days, the cells should be supplemented with 60% fresh DMEM media + M-CSF.
12. On day 9 (1 day before the necroptosis assay), prepare to harvest the cells. Wash the cells with 10 mL PBS and trypsinize with 0.05% trypsin for 5 min at 37 °C.
13. Add 5 mL complete media and gently scrape the cells.
14. Transfer the cells to a new 50 mL tube (replicates can be pooled).
15. Centrifuge the cells at 1000 × *g* for 5 min.
16. Resuspend the cells in 10 mL complete DMEM media, count, plate into appropriate dishes or wells for necroptosis assay (the following day). 50,000 cells/well are plated in a 96-well flat bottom plate 100 μL final volume. The next day (the day of the assay), media is replaced to prepare for the necroptosis assay.

3.1.3 Cell-Based Necroptosis Assay Protocol

1. Cells are plated to prepare for the necroptosis assay. As indicated above, L929 cells are plated the day of the assay and BMDM are split 1 day prior to the necroptosis assay. 50,000 cells/well are plated in a 96-well flat bottom TC plate. Final total volume for the assays is 100 μL.
2. Prepare inhibitor dilutions for a dose-response at an appropriate dose range (including dilution into the final total volume, *see* **Note 4**). For currently available inhibitors where the potency is known, it is recommended to perform the necroptosis

assay with threefold dilutions of doses to achieve a well-defined inhibition curve.

3. Prepare apoptosis inhibitors at appropriate dose including dilution into total final volume (zVAD-FMK for BMDM at 50 μM final dose or QVD-Orph for L929 at 50 μM final dose).
4. Add an appropriate volume of RIP1 kinase inhibitor, apoptosis inhibitor or vehicle. Cells are pretreated with compound and apoptosis inhibitors for 30 min to 1 h.
5. During the incubation, prepare the TNFα at appropriate dose being sure to include dilution into total final volume (BMDM 50 ng/mL TNFα and L929 100 ng/mL TNFα).
6. Add an appropriate volume/well of TNFα or vehicle.
7. Place the lid on plate and perform a quick spin. The plate is incubated for approximately 24 h. A typical range of 19–21 h is sufficient for most assays.
8. Necroptotic cell death induced with TNFα in the presence of the caspase inhibitor zVAD-FMK is evaluated by measuring cellular ATP levels with CellTiter-Glo (CTG) Luminescent Cell Viability assay per manufacturer instructions (*see* **Note 5**). Cell death can also be measured by LDH assay (*see* **Note 6** about the LDH assay).
9. The data from the CTG assay is used to generate a dose-response curve using the appropriate software such as GraphPad Prism software for generating dose–response curves. The data is fitted to a direct response model using a standard I_{max} fit equation as shown in Fig. 1.

3.2 Cellular Selectivity Measures

It is strongly recommended to test for RIP1 cellular selectivity in novel experimental systems. Cellular selectivity can be measured through two assays that have been shown to occur independently of RIP1 kinase activity: TNF-induced activation of NF-κB, and

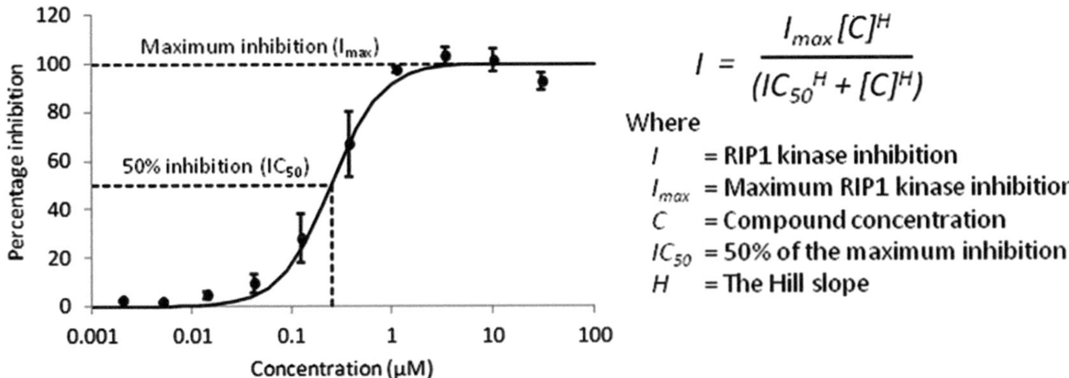

Fig. 1 Representative L929 necroptosis dose response data fitted to a standard I_{max} model (data is the average of two assay runs with $n = 2$ replicates per assay)

TNF + cycloheximide (CHX)-stimulated apoptosis [1, 5]. Again, BMDM cells or L929 cells can be used as controls in this assay. Assay conditions with BMDM are provided as an example.

1. To induce apoptosis with TNF + CHX, BMDM are pretreated with inhibitors or vehicle for 30 min. Cells are then stimulated with TNF (50 ng/mL) and CHX (12 μg/mL). Caspase 3/7 activity is measured at 6 h using the Caspase-Glo 3/7 assay.
2. To assess NF-κB activation, immunoblot analysis is used.
3. For immunoblot analysis, BMDM are pretreated with inhibitors or vehicle for 30 min and then are stimulated with 50 ng/mL TNF for 5 and 15 min.
4. After incubation, lysates are prepared in 1× Cell Lysis Buffer containing protease and phosphatase inhibitors. Cellular lysates are separated on 4–12% SDS-PAGE and blotted onto nitrocellulose membrane. Blots are probed for IκB, phospho-IκB, and loading control.

3.3 Pharmacokinetic and Pharmacodynamic Assays to Determine Dose for In Vivo Disease Models

TNF is a pleiotropic molecule with a crucial role in cellular stress and inflammation during infection, tissue damage, and cancer. RIP1 and RIP3 are key signaling molecules in necrosis and are regulated by caspases and ubiquitination. Moreover, TNF administration in mice results in a sterile shock that is RIP1-dependent. Coadministration of zVAD with TNF induces a more rapid shock that is RIP1 kinase-dependent [1, 5, 6]. Here we will discuss the use of the pharmacokinetic (PK) assay and the TNF/zVAD shock model as a representative acute PD assay to build a pharmacodynamic model for appropriate dose selection of small-molecule inhibitors for use in novel experimental systems where the biology of RIP1 is not well understood.

1. Mice are either bred in house or received from vendor (we typically use C57BL/6 mice). If mice are received from a vendor, then mice are acclimated for at least 1 week prior to use. The recommended age of mice for use in the assay is 8–12 weeks of age.
2. All procedures are performed in accordance with protocols reviewed and approved by an Institutional Animal Care and Use Committee (*see* **Note 7**).

3.3.1 Pharmacokinetic (PK) Sample Collection and Analysis

1. PK analysis allows the determination of how much drug is available in the systemic circulation of the mouse over time. It is important to understand the systemic drug levels over time (PK profile) and ensure proper exposure in experimental models.
2. For PK studies, mice are assigned to treatment groups (n = 3 mice/group) and weighed to ensure proper dosing. A PK study

can be performed with a single dose of compound but including 2–3 doses (separated by tenfold dilutions) will allow assessment of dose linearity.

3. The mice are given a single dose of compound via the appropriate route of administration (*see* **Notes 8** and **9**). Nec-1 and Nec-1s are typically given as an IP dose while our tool molecule GSK'963 can be given as an IP dose or through oral gavage. Once mice are dosed through the appropriate route they are returned to their cages.
4. In order to gain an understanding of the systemic exposures, blood samples are taken at various times post compound administration. For blood collection, we recommend tail tip amputation (*see* **Note 10**). For typical PK studies blood is collected at 15 and 30 min, 1, 2, 4, 6, 8, and 24 h following compound administration.
5. 25 μL of blood is collected in a 25 μL calibrated pipette and placed in a tube and mixed with 25 μL of water and vortexed. The tube is then frozen on dry ice and stored at −80 °C for future DMPK analysis of drug concentrations using liquid chromatography (LC) linked to tandem mass spectral analysis (MS/MS).
6. Prior to bioanalysis, samples are thawed and each analyte is isolated by mixing 3 volumes of acetonitrile with 1 volume of the sample in order to precipitate the protein. The sample is then centrifuged at 950 × *g* for 20 min to pellet the protein before the resulting supernatant is injected onto the LC-MS/MS system optimized for detection of the compound of interest.
7. Data are reported as quantitative drug concentrations as determined by comparison to a standard calibration curve response. Using optimized conditions, a typical lower limit of quantification achieved is around 1.0 ng/mL (Fig. 2).

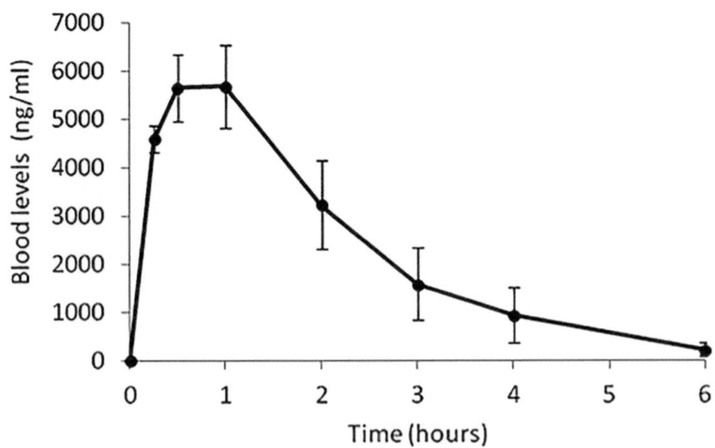

Fig. 2 Representative 30 mg/kg oral mouse ($n = 3$) PK profile (drug concentration in blood versus time)

3.3.2 In Vivo PD Assay (TNF/zVAD-Induced Shock)

1. Mice are assigned to the appropriate experimental groups and weighed to ensure proper dosing. Typical study design is provided with three doses (3–4 doses is recommended to understand in vivo potency).

Group number	Treatment	IV treatment	Number of mice
1	Vehicle	zVAD	3
2	Vehicle	TNFα/zVAD	7
3	RIPK1 inhibitor dose 1	TNFα/zVAD	7
4	RIPK1 inhibitor dose 2	TNFα/zVAD	7
5	RIPK1 inhibitor dose 3	TNFα/zVAD	7

2. Prior to challenge a temperature reading is measured with a rectal thermometer to establish a baseline temperature for each mouse.
3. The mice are administered a single dose of compound or vehicle via an appropriate route 15 min prior to challenge. Mice are then returned to their cages.
4. Fifteen minutes after compound administration, mice are challenged with TNFα and zVAD at 1.25 mg/kg and 16.67 mg/kg, respectively, via intravenous injection (*see* **Note 11**).
5. The mice are then monitored for temperature loss over 3 h (at least every 30 min after the 2 h timepoint). Careful monitoring is needed as the temperature loss can progress very rapidly. A final temperature reading is taken before the mice are euthanized (Fig. 3).
6. A terminal blood sample is collected for PK analysis (on compound treated mice). Additional samples could obviously be taken for other biological assessments (e.g., serum for inflammatory cytokine analysis).
7. In order to generate a dose-response curve, drug levels at the approximate time of the TNF/zVAD challenge for each dose group are needed. This can be obtained from a satellite PK group of 3 mice/dose where blood is collected at 30 min post compound administration. The PK analysis above can be combined with this in vivo PD assay in order to reduce mouse numbers.

3.3.3 Calculation of an In Vivo IC50

1. The blood concentrations at the time of the TNF/zVAD challenge (15 min) can be estimated from the satellite PK mice by comparisons of the terminal measured drug level with a previously determined PK profile. The estimated blood levels for each mouse are plotted against the measured temperature drop.

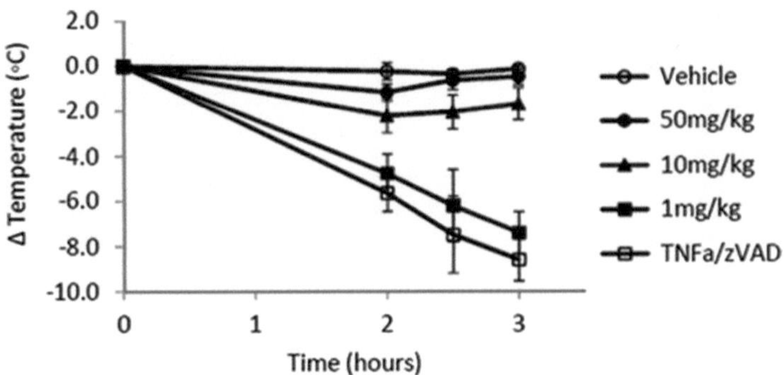

Fig. 3 Representative in vivo PD assay for TNF/zVAD dose response challenge

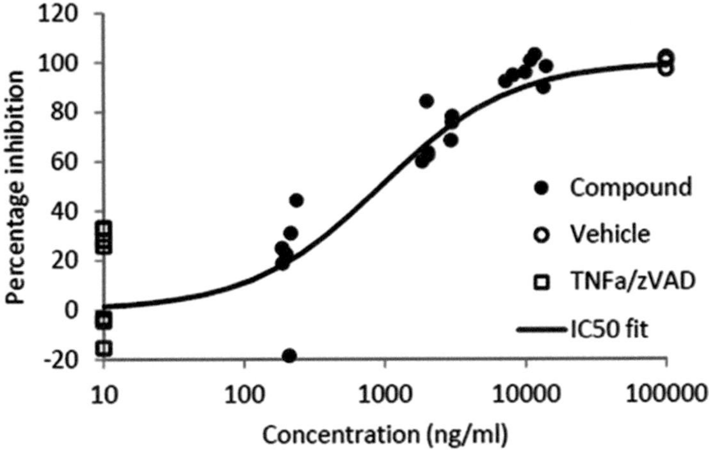

Fig. 4 Concentration versus percentage inhibition of temperature change plot for the TNFa/zVAD model

2. The two control groups are used to define the extremities of the dose response curve, where the vehicle dosed group represents the maximum response to the TNF/zVAD challenge and the non-TNF/zVAD challenged mice represent the minimum response or theoretical maximum level of compound effect.
3. Once a well-defined effect range has been generated where drug concentrations define low to high inhibition levels, the data can be fitted using an I_{max} model (equation shown in Fig. 1). The average temperature change for the vehicle-dosed group defines the no-effect level baseline. The average temperature change in the TNFα/zVAD dosed group defines the maximum effect level. The RIPK1 kinase inhibition achieved for each dose group is then represented as a percentage of the maximum temperature drop in the TNFα/zVAD dosed group (Fig. 4). All standard statistical and graph drawing packages will have a fit option which will return the I_{max} fit characteristics including the IC50 value.

3.3.4 Generating a Pharmacodynamic Model to Select Doses for Disease Models

1. Dose selection for any in vivo model should be based on a targeted drug level to achieve a predefined amount of RIP1 kinase inhibition. This can be to target a level of inhibition at a specific time point (e.g., maximum inhibition at the drug C_{max}) or to maintain inhibition above a level at the drug minimum or even an average level of inhibition over a period of time.
2. Drug concentration time plots should be generated after dosing with the inhibitor via the intended route for the disease model. The drug concentration can then be converted into RIP1 kinase inhibition levels by using the previously determined I_{max} model from Subheading 3.3.3 (Fig. 5).
3. Altering the dosing level (with the assumptions of linear drug pharmacokinetics) will alter the level of inhibition at the various time points and allow the investigator to target different levels of inhibition at different time points (Fig. 6). This allows for

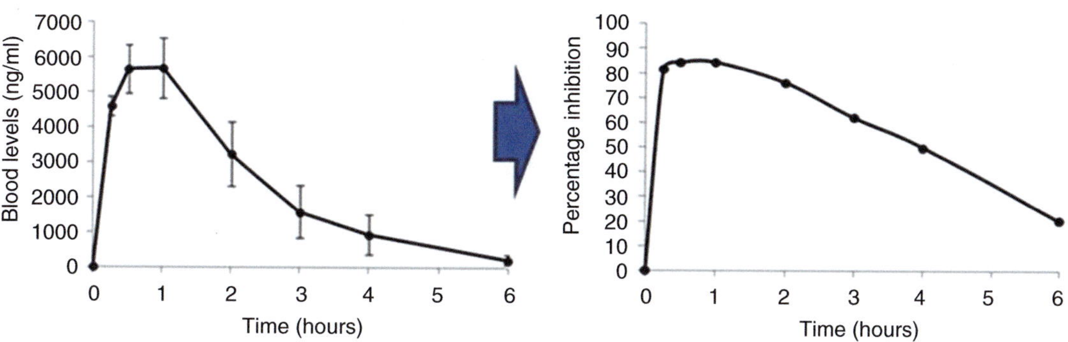

Fig. 5 Conversion of 30 mg/kg po PK profile in mouse into the modelled RIP1 kinase inhibition using the I_{max} model parameters determined from Fig. 4

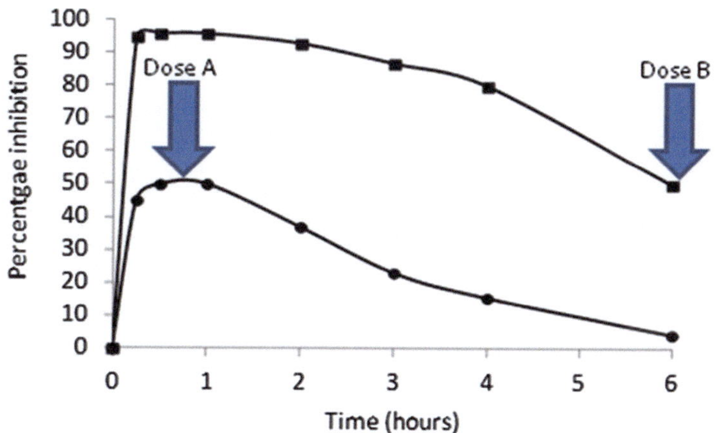

Fig. 6 Scaling of the dose can be used to achieve the desired levels of RIP1 kinase inhibition. Dose A is scaled to 5 mg/kg in order to achieve 50% RIP1 kinase inhibition at the 1 h time point. Dose B is scaled to 130 mg/kg in order to maintain the RIP1 kinase inhibition above 50% over 6 h

the investigation of the relationship between RIP1 kinase inhibition and challenge effects or model readouts.

4. In the case where the RIP1 kinase inhibition is targeted to a specific tissue, further scaling of the drug dose may be required in order to achieve the desired concentrations in that tissue. The scaling factor can be determined by comparing systemic blood levels to the target tissue levels either from direct sampling and measurement by LC-MS/MS or modeling of expected distribution (e.g., through the use of physiological based pharmacokinetic modeling).

4 Notes

1. Media for use in in vitro assays is warmed as needed (too many warm/cold cycles can affect the serum growth factors).
2. Cavitron ((2-hydroxypropyl)-β-cyclodextrin) W7 HP5, Pharma obtained from International Specialty Products (ISP)—Ashland Specialty Ingredients. (2-Hydroxypropyl)-β-cyclodextrin is also available from Sigma.
3. While RIP1 kinase inhibitors can block RIPK1-dependent necroptosis additional information is needed to conclude the involvement of necroptosis in disease since RIP1 regulates cellular processes beyond MLKL-dependent necroptosis. To implicate necroptosis involvement, additional readouts such as assessing necrosome formation, pMLKL, and/or use of MLKL deficient mice is needed [10].
4. When preparing the inhibitors for use in in vitro assays, care is taken to ensure that the DMSO concentration remains as low as possible (less than 1%).
5. For CTG assay, it is important to shake the plate efficiently with a plate shaker to ensure optimal mixing of the CTG reagent. Adherent cells require more mixing to ensure proper lysis.
6. In addition to CTG assay, we also routinely used LDH cell death assay as well. If using the LDH cell death assay, it is recommended to use RPMI-1640 without phenol red. This is important for cell types that have a low signal to noise ratio in the assay. The final volume for the LDH assay is typically 50 μL–100 μL as per manufacturer's instructions.
7. All studies at GSK were conducted in accordance with the GSK Policy on the Care, Welfare and Treatment of Laboratory Animals and were reviewed by the Institutional Animal Care and Use Committee either at GSK or by the ethical review process at the institution where the work was performed.
8. Special note about compound administration. The route of administration depends on the compound. Most tool compounds

can be administered IP. Some compounds can be given through oral gavage, but this depends on the oral exposure. Not all compounds are absorbed well through the oral route. In addition, the exposure of small-molecule inhibitors will vary with the route of administration so it is important to keep the route of administration consistent.

9. For IV dosing the administered materials need to be dissolved and remain in solution and also be filtered through a 0.2 μm filter to ensure suitable sterility. For all other routes, doses may be in a suspension.
10. For tail tip amputation, stabilize the tail with the thumb and forefinger of the hand that will not be used to amputate the very tip of the tail. Using a pair of sharp scissors amputate the tail tip (0.5–1 mm should be adequate, and over time only a maximum of about 2 mm can be removed). Following blood collection ensure good haemostasis with a gauze pad under gentle pressure to the site of amputation. For each collection time point the scab should be removed and the tail gently stroked to collect more blood (25 μL). The bleeding is stopped by applying gentle pressure using a gauze pad.
11. For IV dosing, warming of the tail with a warmed gauze pad that has been dipped in warm water (42 °C) is recommended to help dilate the tail veins. Care is taken to ensure that the temperature of the water is monitored carefully.
12. New tool inhibitors are constantly under development that have similar and improved properties to both GSK'963 and GSK'481. Of particular note these compounds are beginning to demonstrate more favorable properties for use in chronic models.

References

1. Berger SB, Kasparcova V, Hoffman S, Swift B, Dare L, Schaeffer M, Capriotti C, Cook M, Finger J, Hughes-Earle A, Harris PA, Kaiser WJ, Mocarski ES, Bertin J, Gough PJ (2014) Cutting edge: RIP1 kinase activity is dispensable for normal development but is a key regulator of inflammation in SHARPIN-deficient mice. J Immunol 192(12):5476–5480. https://doi.org/10.4049/jimmunol.1400499
2. Kelliher MA, Grimm S, Ishida Y, Kuo F, Stanger BZ, Leder P (1998) The death domain kinase RIP mediates the TNF-induced NF-kappaB signal. Immunity 8(3):297–303
3. Lee TH, Shank J, Cusson N, Kelliher MA (2004) The kinase activity of Rip1 is not required for tumor necrosis factor-alpha-induced IkappaB kinase or p38 MAP kinase activation or for the ubiquitination of Rip1 by Traf2. J Biol Chem 279(32):33185–33191. https://doi.org/10.1074/jbc.M404206200
4. Berger SB, Bertin J, Gough PJ (2015) Drilling into RIP1 biology: what compounds are in your toolkit? Cell Death Dis 6:e1889. https://doi.org/10.1038/cddis.2015.254
5. Berger SB, Harris P, Nagilla R, Kasparcova V, Hoffman S, Swift B, Dare L, Schaeffer M, Capriotti C, Ouellette M, King BW, Wisnoski D, Cox J, Reilly M, Marquis RW, Bertin J, Gough PJ (2015) Characterization of GSK'963: a structurally distinct, potent and selective inhibitor of RIP1 kinase. Cell Death Discov 1:15009. https://doi.org/10.1038/cddiscovery.2015.9

6. Harris PA, King BW, Bandyopadhyay D, Berger SB, Campobasso N, Capriotti CA, Cox JA, Dare L, Dong X, Finger JN, Grady LC, Hoffman SJ, Jeong JU, Kang J, Kasparcova V, Lakdawala AS, Lehr R, McNulty DE, Nagilla R, Ouellette MT, Pao CS, Rendina AR, Schaeffer MC, Summerfield JD, Swift BA, Totoritis RD, Ward P, Zhang A, Zhang D, Marquis RW, Bertin J, Gough PJ (2016) DNA-encoded library screening identifies Benzo[b][1,4]oxazepin-4-ones as highly potent and Monoselective receptor interacting protein 1 kinase inhibitors. J Med Chem 59(5):2163–2178. https://doi.org/10.1021/acs.jmedchem.5b01898
7. Degterev A, Zhou W, Maki JL, Yuan J (2014) Assays for necroptosis and activity of RIP kinases. Methods Enzymol 545:1–33. https://doi.org/10.1016/B978-0-12-801430-1.00001-9
8. He S, Wang L, Miao L, Wang T, Du F, Zhao L, Wang X (2009) Receptor interacting protein kinase-3 determines cellular necrotic response to TNF-alpha. Cell 137(6):1100–1111. https://doi.org/10.1016/j.cell.2009.05.021
9. Sun L, Wang H, Wang Z, He S, Chen S, Liao D, Wang L, Yan J, Liu W, Lei X, Wang X (2012) Mixed lineage kinase domain-like protein mediates necrosis signaling downstream of RIP3 kinase. Cell 148(1–2):213–227. https://doi.org/10.1016/j.cell.2011.11.031
10. Newton K, Dugger DL, Maltzman A, Greve JM, Hedehus M, Martin-McNulty B, Carano RA, Cao TC, van Bruggen N, Bernstein L, Lee WP, Wu X, DeVoss J, Zhang J, Jeet S, Peng I, McKenzie BS, Roose-Girma M, Caplazi P, Diehl L, Webster JD, Vucic D (2016) RIPK3 deficiency or catalytically inactive RIPK1 provides greater benefit than MLKL deficiency in mouse models of inflammation and tissue injury. Cell Death Differ 23(9):1565–1576. https://doi.org/10.1038/cdd.2016.46
11. Degterev A, Hitomi J, Germscheid M, Ch'en IL, Korkina O, Teng X, Abbott D, Cuny GD, Yuan C, Wagner G, Hedrick SM, Gerber SA, Lugovskoy A, Yuan J (2008) Identification of RIP1 kinase as a specific cellular target of necrostatins. Nat Chem Biol 4(5):313–321. https://doi.org/10.1038/nchembio.83
12. Degterev A, Maki JL, Yuan J (2013) Activity and specificity of necrostatin-1, small-molecule inhibitor of RIP1 kinase. Cell Death Differ 20(2):366. https://doi.org/10.1038/cdd.2012.133
13. Muller AJ, DuHadaway JB, Donover PS, Sutanto-Ward E, Prendergast GC (2005) Inhibition of indoleamine 2,3-dioxygenase, an immunoregulatory target of the cancer suppression gene Bin1, potentiates cancer chemotherapy. Nat Med 11(3):312–319. https://doi.org/10.1038/nm1196
14. Teng X, Degterev A, Jagtap P, Xing X, Choi S, Denu R, Yuan J, Cuny GD (2005) Structure-activity relationship study of novel necroptosis inhibitors. Bioorg Med Chem Lett 15(22):5039–5044. https://doi.org/10.1016/j.bmcl.2005.07.077
15. Ren Y, Su Y, Sun L, He S, Meng L, Liao D, Liu X, Ma Y, Liu C, Li S, Ruan H, Lei X, Wang X, Zhang Z (2017) Discovery of a highly potent, selective, and metabolically stable inhibitor of receptor-interacting protein 1 (RIP1) for the treatment of systemic inflammatory response syndrome. J Med Chem 60(3):972–986. https://doi.org/10.1021/acs.jmedchem.6b01196
16. Najjar M, Suebsuwong C, Ray SS, Thapa RJ, Maki JL, Nogusa S, Shah S, Saleh D, Gough PJ, Bertin J, Yuan J, Balachandran S, Cuny GD, Degterev A (2015) Structure guided design of potent and selective ponatinib-based hybrid inhibitors for RIPK1. Cell Rep 10(11):1850–1860. https://doi.org/10.1016/j.celrep.2015.02.052

Chapter 12

Analyzing Necroptosis Using an RIPK1 Kinase Inactive Mouse Model of TNF Shock

Matija Zelic and Michelle A. Kelliher

Abstract

The serine/threonine kinase RIPK1 has numerous biological and pathological functions, mediating pro-survival as well as prodeath apoptotic and necroptotic signaling pathways downstream of various receptors, including death receptors and Toll-like receptors (TLRs). RIPK1 has been implicated in various diseases, including ischemia–reperfusion injury and inflammatory bowel disease (IBD). The recent generation of RIPK1 kinase inactive mice has enabled us to genetically interrogate the role of RIPK1 kinase-mediated necroptosis in disease models. Here, we describe procedures utilizing kinase inactive $Ripk1^{D138N/D138N}$ mice to analyze necroptosis induction in vitro in bone-marrow derived macrophages (BMDMs) and in vivo in a murine model of TNF-induced shock.

Key words Bone marrow-derived macrophages, Lipopolysaccharide, Caspase inhibitor, TNF shock, Cell death, Necroptosis

1 Introduction

Receptor interacting protein kinase 1 (RIPK1) is a serine/threonine kinase crucial in TNF-induced proinflammatory and prosurvival signaling through activation of the NF-κB and MAP kinase pathways. Additionally, RIPK1 can induce apoptotic death or caspase-independent necroptosis, which involves cell and organelle swelling, plasma membrane rupture, and release of damage-associated molecular patterns (DAMPs). RIPK1 and the related RIPK3 auto and transphosphorylate each other, allowing RIPK3 to phosphorylate and activate the pseudokinase MLKL, the downstream executioner of necroptosis and plasma membrane rupture. Necroptosis can be induced downstream of Toll-like receptors (TLR) 3 and 4, or death receptors such as TNFR1 and TRAIL [1, 2]. Although $Ripk1^{-/-}$ mice die at birth [3], the discovery of necrostatin-1 (Nec-1), a small molecule inhibitor of RIPK1 kinase activity and necroptosis [4, 5], has enabled the study of RIPK1 kinase-mediated signaling. However, Nec-1 has off-target effects and a short half-life in vivo [6]. Recently, viable RIPK1 kinase inactive mice were

Adrian T. Ting (ed.), *Programmed Necrosis: Methods and Protocols*, Methods in Molecular Biology, vol. 1857,
https://doi.org/10.1007/978-1-4939-8754-2_12,

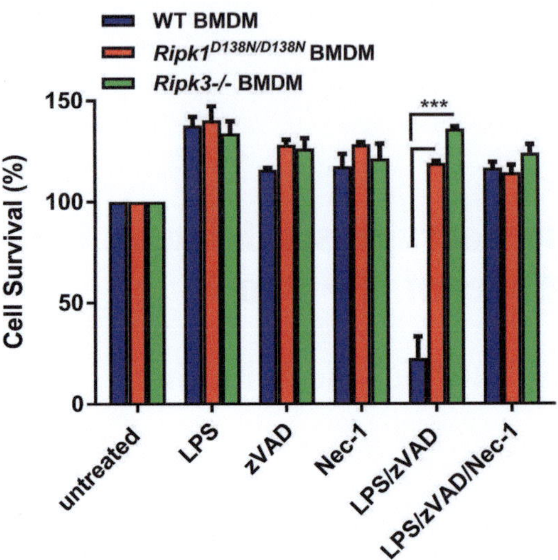

Fig. 1 RIPK1 kinase inactive BMDMs are resistant to LPS/zVAD-induced necroptosis. Primary BMDMs isolated from WT, *Ripk1D138N/D138N* and *Ripk3−/−* mice were treated with zVAD-fmk and/or Nec-1 prior to treatment with LPS. Cell viability was measured with the MTS assay ($n = 3$ mice). Error bars represent SEM. (***$p < 0.001$)

generated; *Ripk1*$^{D138N/D138N}$ targeting the aspartate residue in the activation loop and *Ripk1*$^{K45A/K45A}$ targeting the lysine of the catalytic triad [7–9]. Cells from these mice are resistant to various necroptotic stimuli in vitro and similar to *Ripk3*$^{-/-}$ mice, RIPK1 kinase inactive mice are protected from TNF-induced shock in vivo [7–10]. However, care should be taken in interpreting results from RIPK1 kinase inactive mice since RIPK1 kinase activity can trigger either apoptosis or necroptosis in a cell-type and context-dependent manner. Additionally, RIPK1-independent apoptosis and necroptosis can be triggered [1, 2], and both RIPK1 and RIPK3 kinases can mediate inflammatory signaling independently of necroptotic death, as evidenced in vivo by greater protection of *Ripk1*$^{D138N/D138N}$ and *Ripk3*$^{-/-}$ mice in certain mouse models compared to *Mlkl*$^{-/-}$ mice [11]. Here we describe the isolation and culture of primary bone marrow derived macrophages (BMDMs) isolated from WT, *Ripk1*$^{D138N/D138N}$ and *Ripk3*$^{-/-}$ mice. Macrophages can be induced to undergo RIPK1- and RIPK3-dependent necroptotic death in vitro by inhibiting caspase-8 via a pan-caspase inhibitor, zVAD-fmk, and stimulating TLR signaling by lipopolysaccharide (LPS) addition [12, 13] (Fig. 1). We describe the basics of setting up, optimizing and reading out cell death sensitivity in this in vitro BMDM necroptosis assay. Additionally, we describe using *Ripk1*$^{D138N/D138N}$ mice to analyze necroptotic death in vivo in a mouse model of TNF-induced shock (Fig. 2).

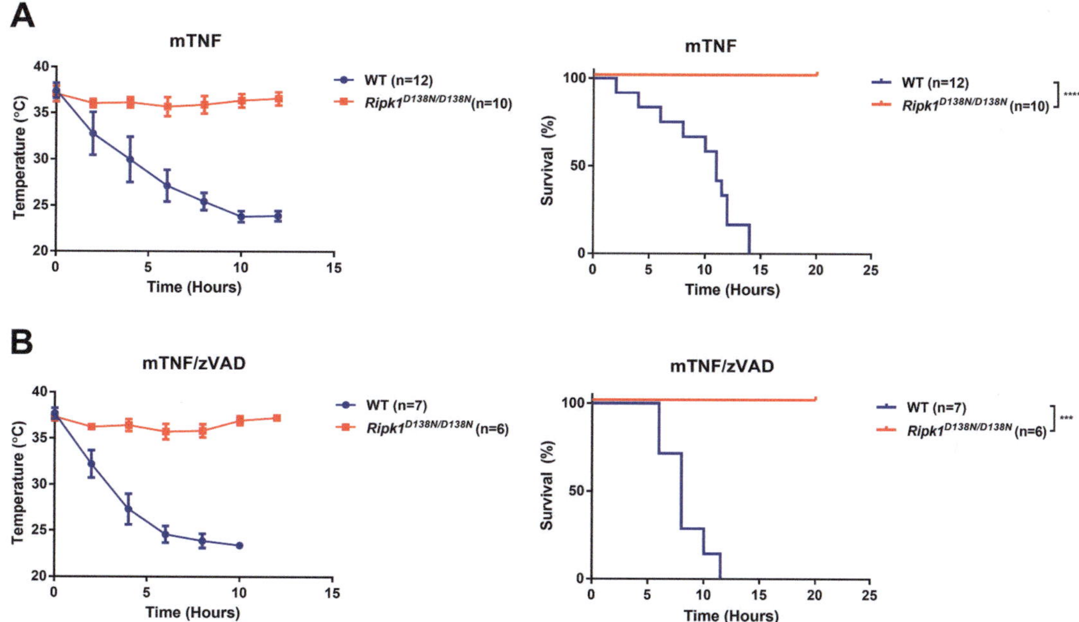

Fig. 2 RIPK1 kinase inactive mice are protected from TNF- and TNF/zVAD-induced shock. Body temperature and survival of WT and *Ripk1D138N/D138N* mice injected with TNF (**a**) or TNF/zVAD (**b**). Error bars represent SEM. (***$p < 0.001$, ****$p < 0.0001$). (Originally published in *The Journal of Immunology*. Polykratis A, Hermance N, Zelic M et al. 2014. Cutting edge: RIPK1 kinase inactive mice are viable and protected from TNF-induced necroptosis in vivo. J Immunol. 193:1539–1543. *Copyright 2014. The American Association of Immunologists, Inc.*)

2 Materials

Prepare all solutions using ultrapure double distilled water (ddH_2O). Store all commercially obtained reagents according to the manufacturer's instructions. Prepare and store all reagents at room temperature, unless otherwise indicated.

2.1 Primary Cells and Culture Conditions (See Chapter 10)

1. Bone marrow derived macrophages (BMDMs) isolated from the bone marrow of WT, *Ripk1*$^{D138N/D138N}$ and *Ripk3*$^{-/-}$ mice.
2. BMDM cell culture media: DMEM, 10% fetal bovine serum (FBS), 1% Penicillin/Streptomycin, 1% L-glutamine, 20% L929 cell supernatant. From a 500 mL bottle of DMEM remove 160 mL, then add 50 mL FBS, 5 mL of Penicillin-Streptomycin, 5 mL L-glutamine, and 100 mL L929 supernatant. Mix well and store at 4 °C. Prior to using the media, warm to 37 °C using a water bath for 10–15 min.
3. 10 cm tissue culture treated dishes.

2.2 Primary Cell Viability Assay

1. Clear flat bottom 96-well tissue culture treated plates.
2. 1.5 mL microcentrifuge tubes, autoclaved.
3. 15 mL conical tubes.

4. Trypan blue (0.4% solution in PBS).
5. Hemocytometer.
6. Dimethyl sulfoxide (DMSO).
7. 0.25% trypsin–EDTA 1× (ThermoFisher).
8. Cell scraper.
9. Phosphate buffered saline (PBS; 10×): 14.4 g Na_2HPO_4, 80 g NaCl, 2 g KCl, 2.4 g KH_2PO_4, ddH_2O to 1 L. Adjust pH to 7.4. Autoclave prior to use. Store at room temperature.
10. Lipopolysaccharide (LPS; 1 mg/mL) (Sigma-Aldrich): Prepare a 1 mg/mL stock solution in sterile-filtered water by weighing 5 mg LPS, then reconstituting in 5 mL water. Mix well, aliquot and store at −20 °C.
11. Necrostatin-1 (Nec-1; 30 mM) (Enzo Life Sciences): Prepare a 30 mM stock solution in DMSO by weighing 10 mg Nec-1 (MW: 259.3) and transfer it into a 1.5 mL microcentrifuge tube. Dissolve the powder in 1.29 mL DMSO to make a 30 mM stock. Mix well, aliquot and store at −20 °C.
12. Z-Val-Ala-Asp-fluoromethylketone (zVAD; 20 mM) (Enzo Life Sciences): Prepare a 20 mM stock solution in DMSO by reconstituting 1 mg zVAD (MW: 467.5) in 107 μL DMSO. Mix well, aliquot and store at −20 °C.
13. Sterile-filtered water (Sigma).
14. CellTiter 96® AQueous One Solution Cell Proliferation Assay (MTS) kit (Promega).
15. Plate reader.

2.3 TNF and TNF/zVAD Shock in Mice

1. Recombinant mouse tumor necrosis factor alpha (rmTNFα; 250 μg/mL) (Cell Sciences): Prepare a 250 μg/mL stock solution in sterile-filtered PBS by reconstituting 1 mg TNFα in 4 mL PBS. Mix well, aliquot and store at −80 °C.
2. zVAD (Bachem; 20 mg/mL): Prepare a 20 mg/mL stock solution in sterile-filtered DMSO by reconstituting 5 mg zVAD in 250 μL DMSO. Mix well and store at −20 °C.
3. Sterile-filtered DMSO (Sigma-Aldrich).
4. Sterile-filtered PBS (Lonza Biowhittaker).
5. Scale.
6. Heat lamp.
7. Mouse restrainer.
8. Digital thermometer.
9. 25G × $^5/_8$ in. needle.
10. 1 mL syringe.
11. WT and *Ripk1*$^{D138N/D138N}$ mice.

3 Methods

Perform all primary cell experiments and reagent reconstitutions and dilutions in an ultraviolet-sterilized vacuum hood. Incubate cells in a 5% CO_2 incubator at 37 °C.

3.1 Primary BMDM Cell Culture

1. Isolation of hematopoietic progenitors for BMDM cultures is carried out as described in Chapter 6. Bone marrow progenitors are cultured in BMDM cell culture media for 7–8 days. Every 2–3 days aspirate all of the medium and loosely adherent cells, wash with sterile 1× PBS, and add 10 mL of fresh prewarmed BMDM cell culture media.
2. If macrophages start to reach confluence around day 4–5, split the cells 1:2. After aspirating the medium and washing with PBS, add 1 mL trypsin and put the plates in the 37 °C incubator for 5 min. Using a cell scraper, gently scrape the adhered macrophages off the bottom of the plate. Add 9 mL prewarmed BMDM cell media and after pipetting multiple times, disperse 5 mL of the cells to each of two new plates. Add 5 mL fresh BMDM cell media to each plate, swirl the plates well to make sure that the cells spread out evenly, and continue incubating at 37 °C.

3.2 Plating BMDMs for Viability Assay

1. On day 7 or 8, collect the BMDMs with trypsin and gentle scraping as described in Subheading 3.1, **step 2**.
2. Add prewarmed BMDM cell culture media, pipet multiple times to collect the cells, transfer to a 15 mL conical tube and centrifuge at 440 × *g* for 5 min to pellet the cells.
3. Aspirate the media and resuspend the BMDM cell pellet in 5 mL BMDM cell culture media to count the cells with a hemocytometer. Briefly, swirl the flask and take out 20 μL of the cells into a 1.5 mL microcentrifuge tube. Mix with 80 μL 0.4% trypan blue and load 10 μL onto the hemocytometer. Count the cells and determine the cell number per mL using the following formula (*see* **Notes 1** and **2**):

   Number of cells per mL = [number of cells in four corner squares]/4 × dilution factor × 10^4
4. Plate 5 × 10^4 BMDMs in 100 μL per well in triplicate in 96-well plates for each genotype and treatment condition that will be used (*see* **Note 3**). Here there are three genotypes (WT, $Ripk1^{D138N/D138N}$, and $Ripk3^{-/-}$) and six treatment conditions (untreated, LPS, zVAD, Nec-1, LPS/zVAD, and LPS/zVAD/Nec-1) (*see* **Note 4**).
5. Plate the cells in BMDM cell culture media and incubate the plates at 37 °C for 3–4 h to let the BMDMs adhere.

3.3 Necroptotic Viability Assay In Vitro

1. Once the cells have adhered, aspirate the media and gently wash the wells with 1× PBS. Add 100 μL DMEM/10% FBS media to all the wells with cells, omitting the L929 supernatant (*see* **Note 5**).
2. Pretreat the cells for 1 h with zVAD (20 μM final concentration) and/or Nec-1 (30 μM final concentration) and incubate at 37 °C (*see* **Note 6**).
3. Add LPS to the cells (20 ng/mL final concentration) and incubate at 37 °C for 20 h (*see* **Note 7**).
4. Add 20 μL of CellTiter 96® AQueous One Solution Cell Proliferation Assay (MTS) reagent to each well (*see* **Note 8**), mix by pipetting, and incubate at 37 °C for at least 1 h (*see* **Note 9**). Record the absorbance at 490 nm using a spectrophotometer (Fig. 1) (*see* **Note 10**).
5. To determine the relative cell viability in response to each treatment condition, first subtract the average of the blank well triplicate (media only) from each treatment absorbance reading, then average the triplicate values for each treatment condition. Utilize the following formula to obtain the relative percent cell viability of each individual treatment group compared to untreated control cells for the respective genotype (*see* **Note 11**):

Relative % viability = ([average A_{490} treated cell]/[average A_{490} untreated cell]) × 100

3.4 TNF and TNF/zVAD Shock In Vivo

1. For TNF-induced shock studies it is best to use age- and sex-matched mice, ideally littermates between the ages of 6–12 weeks (*see* **Note 12**).
2. To calculate an average weight use a scale to weigh the mice that will be used for the experiment (*see* **Note 13**).
3. Based on the mouse number and average weight, make up dilutions of rmTNFα and zVAD in sterile-filtered PBS (*see* **Note 14**). For TNF-induced shock inject 9 μg TNF per mouse in 200 μL PBS. For TNF/zVAD-induced shock inject 9 μg TNF and 16.7 mg/kg zVAD in 200 μL PBS. For example, a 25 g average weight would require 36 μL TNF, 20.9 μL zVAD, and 143.1 μL PBS per mouse (*see* **Note 15**).
4. Warm the mice up for 5 min by placing the cage next to a heat lamp (*see* **Note 16**).
5. Place a detachable 25G × $^5/_8$ in. needle on a 1 mL syringe, making sure that the bevel faces up when aligned with the barrel measurements. Draw up 200 μL of the TNF or TNF–zVAD mixture in PBS for each mouse, making sure to get rid of any air bubbles in the syringe.
6. Place the prewarmed mouse in a restrainer, spray down the tail with 70% ethanol, and inject the TNF or TNF–zVAD mixture intravenously into the tail vein (*see* **Note 17**).

7. Once all the mice have been injected, monitor their temperature every few hours. Use the digital thermometer to record the temperature rectally (*see* **Note 18**).
8. Euthanize moribund mice and record time of sacrifice/death (Fig. 2) (*see* **Note 19**).

4 Notes

1. Count the live unstained cells, do not count blue stained cells as these are not viable.
2. In this case, the dilution factor is 5 (20 μL cells mixed with 80 μL trypan blue in a total volume of 100 μL) so input 5 into the equation to obtain the number of cells per mL.
3. With the MTS assay, MTS conversion results in a yellow to purple/dark brown color change, which is proportional to the number of living cells in culture. Cell growth rate and under-confluence or overconfluence can affect the accuracy of the results, thus it is best to optimize the cell number based on the cell type being used. With primary BMDMs, a good range to test would be 2×10^4, 5×10^4, and 1×10^5 cells per well in a 96-well plate. Cells can also be plated over-night, instead of for 3–4 h before starting treatment.
4. It is best to always take more cells than will be necessary, in this case 1×10^6 cells would be sufficient for 20 wells.
5. MTS assays can have a high background reading, so make sure to have three control wells with no cells and 100 μL media per well.
6. For both zVAD and Nec-1 the final dilution is 1:1000 from the stock concentration. Preferably, create a master mix of zVAD and/or Nec-1 in media at the final concentration, and add 100 μL to each well as needed, after having aspirated and washed the old media. Alternatively, dilute each compound 1:10 in media, then add 1 μL per well into 100 μL media.
7. Dilute the LPS 1:1000 in media to a concentration of 1 μg/mL, then add 2 μL per well. Alternatively, make master mixes containing LPS, with or without zVAD and Nec-1, diluted to the final concentration and add 100 μL per well after aspirating and washing the old media.
8. The MTS reagent is light sensitive, so make sure to protect it from light. Thaw the reagent at room temperature. Open the bottle and add MTS to the plate in a sterilized vacuum hood with the light off, to prevent contamination and light damage.
9. With the MTS assay it is necessary to test a series of incubation times (e.g., 30 min, 1 h, 2 h, 3 h, etc.) after adding the MTS reagent. It is important to determine an optimal time so that

the color signal is not saturated. The absorbance reading should be in the range from 0.0 to 2.0.

10. LPS, zVAD or Nec-1 should not induce death in BMDMs on their own. In a preliminary experiment, test a range of concentrations for each reagent to find the optimum tolerated concentration range. In combination, LPS and zVAD should induce death in WT BMDMs, which should be rescued with addition of Nec-1. Concentrations of reagents may also need to be optimized to see these effects.

11. RIPK1 kinase activity can mediate both apoptotic and necroptotic death, and RIPK1-independent necroptosis can occur [1, 2], so additional controls in addition to WT and *Ripk1*$^{D138N/D138N}$ BMDMs should be utilized. Here, Nec-1 and *Ripk3*$^{-/-}$ BMDMs are added to test necroptotic death. *Mlkl*$^{-/-}$ BMDMs or RIPK3 kinase inhibitors can also be used. Furthermore, the MTS death assay can be utilized with other cell types isolated from RIPK1 kinase inactive mice, including mouse embryonic fibroblasts (MEFs) or bone marrow derived dendritic cells (BMDCs). These cells can be tested for necroptosis susceptibility with TNF/Cycloheximide/zVAD or TNF/Smac mimetic/zVAD treatment. Finally, other viability assays including CellTiter-Glo® and flow cytometry-based annexin V/propidium iodide staining can be used instead of the MTS assay.

12. Gender differences in susceptibility to shock and sepsis exist in both mouse models and patients. In general, females are thought to be more resistant due to decreased production of proinflammatory cytokines and increased production of immunosuppressive mediators [14, 15].

    Additionally, differences in the gut microbiome can affect result interpretation in many mouse models [16]. Thus, it is important to use age- and sex-matched littermates in these TNF-induced shock experiments. If littermates are not available, it is best to wean mice together into the same cage or mix the bedding between cages for several weeks before performing the experiment.

13. If possible it is best to use mice of a similar weight, or a similar weight range for the different genotypes, in this case WT and *Ripk1*$^{D138N/D138N}$ mice.

14. Thaw the TNF and zVAD on ice, and keep on ice once you have prepared the mixture.

15. Make a master mix based on the experimental mouse number. Remember to always make enough for a couple extra mice, as ~100 μL volume is lost in the shaft of the needle.

16. Warming up the mice helps to dilate the vein and makes it easier to visualize.

17. Inject at a shallow angle with the bevel of the needle pointed up, since the vein is close to the tail surface. Start injecting closer to the bottom half of the tail, so there is ample room to move up the tail if the initial injection angle is not successful. Push lightly on the plunger, if there is any resistance the needle is not in the vein. Press on the injection site briefly with a paper towel to stop the bleeding, and place the mouse back into the cage for recovery and monitoring.
18. Use sterile autoclaved Vaseline on the tip of the thermometer probe, and clean it between measurements on each mouse.
19. Before starting the shock experiments, obtain institutional approval from the Animal Care and Use Committee (IACUC) and follow the guidelines for monitoring and euthanasia as approved by the institution.

Acknowledgments

This work was supported by National Institutes of Health/National Institute of Allergy and Infectious Diseases Grant AI075118.

References

1. Pasparakis M, Vandenabeele P (2015) Necroptosis and its role in inflammation. Nature 517:311–320. https://doi.org/10.1038/nature14191
2. Weinlich R, Oberst A, Beere HM et al (2017) Necroptosis in development, inflammation and disease. Nat Rev Mol Cell Biol 18:127–136. https://doi.org/10.1038/nrm.2016.149
3. Kelliher MA, Grimm S, Ishida Y et al (1998) The death domain kinase RIP mediates the TNF-induced NF-kappaB signal. Immunity 8:297–303
4. Degterev A, Huang Z, Boyce M et al (2005) Chemical inhibitor of nonapoptotic cell death with therapeutic potential for ischemic brain injury. Nat Chem Biol 1:112–119
5. Degterev A, Hitomi J, Germscheid M et al (2008) Identification of RIP1 kinase as a specific cellular target of necrostatins. Nat Chem Biol 4:313–321. https://doi.org/10.1038/nchembio.83
6. Takahashi N, Duprez L, Grootjans S et al (2012) Necrostatin-1 analogues: critical issues on the specificity, activity and in vivo use in experimental disease models. Cell Death Dis 3:e437. https://doi.org/10.1038/cddis.2012.176
7. Newton K, Dugger DL, Wickliffe KE et al (2014) Activity of protein kinase RIPK3 determines whether cells die by necroptosis or apoptosis. Science 343:1357–1360. https://doi.org/10.1126/science.1249361
8. Polykratis A, Hermance N, Zelic M et al (2014) Cutting edge: RIPK1 kinase inactive mice are viable and protected from TNF-induced necroptosis in vivo. J Immunol 193:1539–1543. https://doi.org/10.4049/jimmunol.1400590
9. Berger SB, Kasparcova V, Hoffman S et al (2014) Cutting edge: RIP1 kinase activity is dispensable for normal development but is a key regulator of inflammation in SHARPIN-deficient mice. J Immunol 192:5476–5480. https://doi.org/10.4049/jimmunol.1400499
10. Duprez L, Takahashi N, Van Hauwermeiren F et al (2011) RIP kinase-dependent necrosis drives lethal systemic inflammatory response syndrome. Immunity 35:908–918. https://doi.org/10.1016/j.immuni.2011.09.020
11. Newton K, Dugger DL, Maltzman A et al (2016) RIPK3 deficiency or catalytically inactive RIPK1 provides greater benefit than MLKL deficiency in mouse models of inflammation and tissue injury. Cell Death Differ 23:1565–1576. https://doi.org/10.1038/cdd.2016.46
12. He S, Liang Y, Shao F et al (2011) Toll-like receptors activate programmed necrosis in

macrophages through a receptor-interacting kinase-3-mediated pathway. Proc Natl Acad Sci U S A 108:20054–20059. https://doi.org/10.1073/pnas.1116302108

13. Kaiser WJ, Sridharan H, Huang C et al (2013) Toll-like receptor 3-mediated necrosis via TRIF, RIP3, and MLKL. J Biol Chem 288:31268–31279. https://doi.org/10.1074/jbc.M113.462341
14. Marriott I, Huet-Hudson YM (2006) Sexual dimorphism in innate immune responses to infectious organisms. Immunol Res 34:177–192
15. Angele MK, Pratschke S, Hubbard WJ et al (2014) Gender differences in sepsis: cardiovascular and immunological aspects. Virulence 5:12–19. https://doi.org/10.4161/viru.26982
16. Laukens D, Brinkman BM, Raes J et al (2016) Heterogeneity of the gut microbiome in mice: guidelines for optimizing experimental design. FEMS Microbiol Rev 40:117–132. https://doi.org/10.1093/femsre/fuv036

Chapter 13

Assessment of In Vivo Kidney Cell Death: Acute Kidney Injury

Wulf Tonnus, Moath Al-Mekhlafi, Christian Hugo, and Andreas Linkermann

Abstract

The kidney has been studied as an organ to investigate cell death in vivo for a number of reasons. The unique vasculature that does not contain collateral vessels favors the kidney over other organs for the investigation of ischemia–reperfusion injury. Unilateral uretic obstruction has become the most prominently studied model for fibrosis with impact far beyond postrenal kidney injury. In addition, the tubular elimination mechanisms render the kidney susceptible to toxicity models, such as cisplatin-induced acute kidney injury. During trauma of skeletal muscles, myoglobulin deposition causes tubular cell death in the model of rhabdomyolysis-induced acute kidney injury. Here, we introduce these clinically relevant in vivo models of acute kidney injury (AKI) and critically review the protocols we use to effectively induce them.

Key words Acute kidney injury (AKI), Kidney ischemia–reperfusion injury, Unilateral ureteric obstruction, Cisplatin-induced AKI, Rhabdomyolysis-induced AKI

1 Kidney Ischemia–Reperfusion Injury

1.1 Introduction

Kidney ischemia–reperfusion injury (hereafter: IRI) is a well-established model of transient ischemia to a whole organ. This widely used prototype model of acute kidney injury (AKI) closely phenycopies solid organ transplantation and hypoxia during prerenal AKI. Necrosis of proximal tubules is a pathologic hallmark of murine ischemic kidney injury [1], which renders this model highly relevant for research on regulated necrosis. Indeed, significant knowledge about necroptosis and ferroptosis in vivo has been gained from this model [2–4]. IRI is induced by transient clamping of both renal arteries. Two approaches are frequently followed, (1) from the back of the mouse, which is less traumatic, (2) transabdominal surgery which we prefer due to a better view of the operating area which has led to less standard deviation and higher precision. Alternative approaches include unilateral nephrectomy and contralateral clamping [5], but in our hands, these models did

Adrian T. Ting (ed.), *Programmed Necrosis: Methods and Protocols*, Methods in Molecular Biology, vol. 1857,
https://doi.org/10.1007/978-1-4939-8754-2_13,

not yield reproducible results for measures of serum creatinine and serum urea concentrations as markers of renal function and intoxication, respectively. Instead, this model was associated with high standard deviations in our hands. Here, we therefore share our protocol of abdominal bilateral kidney IRI. Typical readouts for this model are survival time/rate, histopathological features, protein and RNA quantification, and serum parameters of AKI.

1.2 Materials

1.2.1 Technical Equipment

1. A vaporizer for inhalative anesthetics (isoflurane) (*see* **Note 1**).
2. An autoregulated heating plate (36 °C and 37 °C) (*see* **Note 2**).
3. Surgical instruments: A needle-holder, precision scissors, anatomic forceps, pointed forceps, larger forceps, 6-0 sutures.
4. A surgical microscope and a small-animal surgery unit including magnetic retractors.

1.2.2 Other Materials

1. A set of two microaneurysm clamps (*see* **Note 3**).
2. A 20 mL syringe of prewarmed PBS must be available at all time during surgery (*see* **Note 4**).
3. An analgesic with favourable pharmacokinetics, e.g., buprenorphine–HCl (*see* **Note 5**).
4. Optional: prepared, prewarmed compound, e.g., 1.65 mg/kg body weight Nec-1.
5. Prepare a couple of 2 × 2 cm clinically available compresses.

1.3 Methods

There are no special requirements for the surgical room apart from constant room temperature.

1. At least six cohoused littermates, which are carefully matched for size and weight, should be used per group (*see* **Note 6**). These mice should be between 8 and 10 weeks of age. If the standard deviation in the control goup is higher than 0.2 mg/dL serum creatinine, the number shoud be increased to 8–10 mice per group.
2. 15 min prior to the onset of surgery, mice receive an injection of the analgesic i.m. into the right hind limb (*see* **Note 7**), prewarmed investigated compounds may be injected in a maximal volume of no more than 150 μL i.p. 15 min before the onset of surgery. Control mice receive the same amount of prewarmed PBS to avoid unequal fluid load, which is critical in the analysis of all renal parameters.
3. Inhalation narcosis should be induced at 3–5 L/min isoflurane (*see* **Note 8**).
4. The anesthetized mouse should be quickly placed (on its back) onto the autoregulatory heating plate and narcosis should be maintained by placing the mouse's nose in the narcosis tube.

5. The mouse should be immobilized with a fixing stripe at each extremity.
6. The abdominal fur is removed by the method of choice (e.g., with a gel or an electric shaver).
7. Use the precision scissors to open the cutis and peritoneum parietale layer-by-layer, resulting in an approx. 20 mm opening (*see* **Note 9**).
8. Place a retractor on the left and right side each to spread the opening wound.
9. Rinse a prepared compress in the PBS falcon tube and place it on the left side of the wound.
10. Carefully identify the caecum and mobilize the intestine. Carefully pull the intestine out of the abdomen to the mouses' left side and cover it with a second compress.
11. Place the third retractor at 11 o'clock to get view on the right renal artery (*see* **Note 10**).
12. Carefully open the retroperitoneum with the pointed forceps and place the microaneurysm clamp on the renal pedicle and immediately start the ischemia timer (*see* **Note 3**). Do not clamp the artery or the vein alone.
13. Quickly remove the 11 o'clock retractor, switch the intestine to the other side and replace the retractor at 1 o'clock without touching the pancreas.
14. Repeat **step 12** (*see* **Note 11**), note the seconds between both clamps.
15. Take away all retractors, replace the intestine in the abdomen and cover the opening with the used dressings (*see* **Note 12**).
16. 1 min before your ischemia timer ends, repeat **steps 8–12** to regain excess to the clamp on the right renal artery, remove it by the second the ischemic timer ends (*see* **Note 13**). We prefer to remove the clip at the particular second (!). Mice that are for any reason left on the surgery board for more than 10 s over time are removed from the experiment.
17. Repeat **step 14**, then remove the left clamp at the interval you had at placing them.
18. Confirm reperfusion by visual control of the changing color of the kidneys.
19. Replace the intestine into the abdomen.
20. Close the peritoneum parietale and the cutis with continuous 6-0 sutures in two separate layers. Do not connect the peritoneum to the cutis (*see* **Note 14**).
21. Stop anesthesia, immediately inject 1 mL prewarmed PBS i.p. and place the mouse in its cage.

22. Within less then 15 min, very often within 5 min, the mouse will start to move in the cage without much handicap (*see* **Note 15**).
23. After 24 h or 48 h, samples can be collected (*see* **Notes 16–18**).

1.4 Notes

1. Volatile anesthetics are superior to for example ketamine injections because of easier pharmacokinetics and less effects on blood pressure.
2. The lower abdominal region should be placed at the sensor of the autoregulatory heating system. The sensor must not be introduced rectally but has to be fixed on top of the heating pad right underneath the lower mouse abdomen.
3. These clamps are extraordinary sensitive. If pressure is not homogenous, high standard deviations will result. Such clamps should be replaced immediately. We place the base of the clip on the renal pedicle, not the pinch. Even with the very best clamps available, we routinely change clamps after no more than 500 IRI procedures.
4. Prewarming can be achieved by any system. Placing a full syringe on a heating sterilizer may be sufficient.
5. Buprenorphine is an opioid with a long half-life. If mice show signs of pain after surgery, you should give another shot to reduce stress of the animal. Make sure to follow regional laws regarding opioid use and storage.
6. Only use male mice for IRI, as there is an immunologic component, which is affected by female hormone levels. We recommend not placing more than four animals into one cage.
7. Opioid analgesics should preferentially be given intraperitoneally rather than intramuscularly. However, if you use another compound intraperitoneally, interaction might happen, so intramuscular injection is superior in that scenario.
8. 3–5 L/min of isoflurane airflow has proven to be an adequate dose for induction of narcosis. For maintainance during surgery and ischemia time, we recommend to reduce this dose by 50%. Otherwise, circulation is suppressed and mice might die during surgery from overdosage.
9. For the first opening in each layer, lift the layer with the forceps and cut a small opening by twisting the scissors 90° to the latter curring direction. Twist the scissors by 90° and cut the linea alba without opening the peritoneum. When opening the peritoneum parietale with a new onset cut, prevent injury to the intestine.
10. The liver covers the renal artery after a midline incision. To move the liver aside, two techniques are recommended: (1)

carefully use the retractor to pull the liver as well or (2) fix a moisture Q-tip with your needle holder and carefully prevent the liver from falling back into the peritoneum with the rinsed tip. The same applies for the spleen on the left side, which, especially in knockout animals, can be quite large.

11. Importantly, the interval between both clamps shall be less than 90 s.
12. When covering the wound, rinse the surface with prewarmed PBS to improve fluid balance and temperature control.
13. Ischemic tolerance can be different between mice. For male C57Bl/6 N mice, 30 min results in severe nonlethal IRI.
14. Use a continuous 6-0 suture for the peritoneum. Cut the endings short to prevent the mouse from opening the suture.
15. Upon postsurgery recovery (15 min), allow no more than two mice to share a cage.
16. We use the same anesthesia protocol for blood collection from the retro-orbital venous plexus.
17. After blood collection, sacrifice mice and harvest kidneys.
18. Double-blinding can further improve reliability: an assistant might hand over the mice to the surgeon. Also, for histopathological evaluation, blinding is inevitable.

2 Unilateral Ureteric Obstruction (UUO)

2.1 Introduction

Unilateral ureteric obstruction (hereafter: UUO) is a common model of postrenal kidney failure as well as induction of end-stage renal disease with fibrosis, but it is most commonly used as a general model for fibrosis. It is currently unclear which cell death pathways are involved in the process. Obstruction is achieved by ligation of the left ureter with surgical knots or, alternatively, with clamps. Here, we provide our protocol on the ligation with knots. Readouts of this model typically focus on histopathological aspects, as the most reliable serum parameters are unaffected because of compensatory function of the contralateral kidney.

2.2 Materials

2.2.1 Technical Equipment

The technical requirements are the same as for Subheading 1.2.1—**IRI**.

2.2.2 Other Materials

1. 6-0 sutures for ureteric ligation.
2. A falcon tube with fresh PBS (room temperature) at hand.
3. A 20 mL syringe of prewarmed PBS (*see* **Note 4** in Subheading 1.4).

4. An analgetic with favorable pharmacokinetics, e.g., buprenorphine–HCl (*see* **Note 5** in Subheading 1.4).
5. Optional: your prepared, prewarmed compound, e.g., 1.65 mg/kg body weight Nec-1.
6. Prepare a couple of 2 × 2 cm clinically available compresses.

2.3 Method

There are no special requirements for the surgical room. Unlike IRI, temperature is less critical for this model.

1. At least six cohoused littermates at the age of 8–10 weeks, carefully matched for size and weight, should be used per group (*see* **Note 6** of Subheading 1.4).
2. Fifteen minutes prior to the onset of inhalation narcosis, mice receive an injection of the analgetic intramuscularly (*see* **Note 7** of Subheading 1.4), prewarmed investigated compunds may be injected in no more than 150 μL intraperitoneally.
3. Inhalation narcosis should be induced with 3–4 L/min isoflurane (*see* **Note 8** of Subheading 1.4).
4. The anesthetized mouse should be quickly placed (on their backside) on the autoregulatory heating plate and narcosis should be maintained by placing the mouse's nose in the narcosis tube.
5. The mouse should be immobilized with a fixing strip at each extremity.
6. The abdominal fur is removed by the method of choice (e.g., with a gel or an electric shaver).
7. For the first opening in each layer, lift the layer with the forceps and cut a small opening by twisting the scissors 90° to the later cutting direction. Twist the scissors 90° and cut the linea alba without opening the peritoneum. When opening the peritoneum parietale with a new onset cut, prevent injury to the intestine. Place a retractor on the left and right side each to spread the opening. Rinse a prepared compress with PBS and place it on the right side of the mouse (*see* **Note 1** of Subheading 2.4).
8. Mobilize the gut and pull it out of the abdomen onto the compress, then cover it with another one to prevent drying (*see* **Note 2** of Subheading 2.4).
9. Identify the left ureter (*see* **Notes 3–5** of Subheading 2.4) under the peritoneal layer—take a surgical microscope if needed.
10. Use the pointed forceps to open the retroperitoneum laterally of the ureter (*see* **Note 6** of Subheading 2.4).
11. Place two surgical double-knots on the ureter (*see* **Notes 7** and **8** of Subheading 2.4).
12. Replace the intestine in the abdominal cave.
13. Close the peritoneum parietale and the cutis with two separate layers of continuous 6-0 sutures (*see* **Note 14** of Subheading 1.4).

14. Stop anesthesia and inject 1 mL prewarmed PBS i.p.
15. Within less then 5 min, the mouse should be moving (*see* **Note 15** of Subheading 1.4).
16. After a prespecified time, samples can be collected (*see* **Notes 9** and **10** of Subheading 2.4). For most purposes, ligation times range between 1 and 2 weeks (*see* **Notes 16** and **17** of Subheading 1.4). However, shorter ligation times should be considered when studying cell death in the early phase of the disease.

2.4 Notes

1. For intestinal mobilization, some find it more convenient to place the dressing on the left side and, after complete mobilization, switch the "package" to the right side.
2. Start to mobilize the intestine by identifying the caecum. It is the "free part" of the intestine and, thus, a safe starter.
3. The left (rather then the right) ureter is used, as there is more anatomical room on this side.
4. It can be helpful to "asymmetrically" fixate the mouse slightly to their right side, so the intestine is removed naturally.
5. To identify the ureter on the *M. ileopsoas*, one should look for a small artery on its abdominal surface.
6. On the medial side of the ureter, arteries such as the aorta abdominalis are located, so one should be very careful with sharp instruments in this field.
7. Two double-knots provide more safety to optimally ligate the ureter. For this purpose, one set of knots should be placed proximally (closely to the kidney) and one set rather distally (direction bladder).
8. For control mice, one should perform the whole procedure similarly but not tie the knots.
9. When using surgical instruments in the mouse, they should be moistened by dipping them in the fresh PBS tube to reduce friction and accidental harm.
10. Double-blinding can further improve reliability: an assistant might hand over the mice to the surgeon, who does not know, which mouse or compound is used. Also, for histopathological evaluation, blinding is helpful.

3 Cisplatin-Induced Acute Kidney Injury

3.1 Introduction

Cisplatin is a chemotherapeutic agent commonly used in combination for epithelial malignancies including testicular and lung cancers [6]. Renal dysfunction is a common adverse event of cisplatin therapy, affecting almost 30% of patients [7]. In mice, a single shot of 20 mg/kg body weight cisplatin reproducibly induces renal injury. Necroptosis and apoptosis are clearly involved in this process, but other cell death pathways have hardly been investigated [8, 9]. This model is simple to establish.

3.2 Materials

1. Fresh cisplatin, prepared in PBS at a final concentration of 1 mg/mL.
2. Optional: your prepared, prewarmed compound, e.g., Nec-1.

3.3 Method

1. At least eight cohoused littermates, which are carefully matched for size and weight, should be used per group (*see* **Note 6** of Subheading 1.4). These mice should be between 8 and 10 weeks of age. We prefer to use male mice, which are slightly more sensitive to renal injury.
2. Optional: Inject the compound to be tested 15 min prior to cisplatin intraperitoneally, up to 150 μL (*see* **Note 1** of Subheading 3.4).
3. Inject 20 mg/kg body weight cisplatin i.p. (*see* **Note 2** of Subheading 3.4).
4. After 48 h, blood and kidney samples are harvested, unless a survival experiment is planned.

3.4 Notes

1. Do not inject compound and cisplatin together, as this delays resorption of both.
2. Only use freshly prepared cisplatin, as chemotherapeutics for humans are prepared on demand as well. Cisplatin is light sensitive and should be stored in the dark.

4 Rhabdomyolysis-Induced Acute Kidney Injury

4.1 Introduction

Rhabdomyolysis-induced acute kidney injury is induced by controlled destruction of a muscle by glycerol injection [10]. This mimics the clinical phenomenon of rhabdomyolysis [11]. The exact pathomechanism of acute kidney injury in rhabdomyolysis remains unknown, but tubular necrosis is a recognized histopathological hallmark [12, 13]. Typical readouts are survival time, histopathological damage, and serum parameters such as creatinine, BUN, and creatinine kinase (CK). Although it is easy to establish, standardization of this model can be tricky regarding constant muscular damage.

4.2 Materials

1. Prepare 50% glycerol solution for injection.
2. A 50 mL falcon tube with fresh PBS (room temperature) should be prepared.
3. An analgesic with favorable pharmacokinetics, e.g., buprenorphine–HCl (*see* **Note 5** of Subheading 1.4).
4. Optional: your prepared, prewarmed compound, e.g., Nec-1.

4.3 Method

1. At least ten cohoused littermates, which are carefully matched for size and weight, should be used per group (*see* **Note 6** of Subheading 1.4). These mice should be between 8 and 10 weeks of age.
2. Initiate water deprivation 12 h before glycerol injection (*see* **Note 1** of Subheading 4.4). A schematic of the protocol is shown in Fig. 1.
3. After 12 h of water deprivation, start with induction of inhalation narcosis.
4. Inhalation narcosis should be induced at 3–5 L/min isoflurane (*see* **Note 8** of Subheading 1.4).
5. After onset of inhalation narcosis, mice receive an injection of an appropriate analgesic (e.g., buprenorphine–HCl) intraperitoneally. At this time point, prewarmed compounds may also be injected i.p. in no more than 150 μL as well.
6. The anesthesized mice should lie on their backside.
7. Rhabdomyolysis is induced by injection of 5 μL glycerol per gram body weight in the right hind limb.
8. Place mice in their cages and let them recover from anesthesia.
9. Water supply should be inhibited for another 12 h before water supply can be allowed ad libitum (*see* **Note 1** of Subheading 4.4).
10. 24–48 h after rhabdomyolysis induction, samples can be collected.

4.4 Notes

1. To investigate stronger kidney injury, water deprivation can be applied up to 24 h past rhabdomyolysis induction.

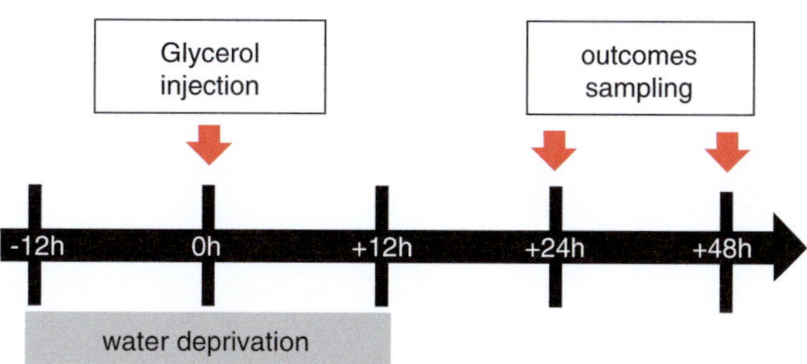

Fig. 1 Timeline of rhabdomyolysis induction: 12 h before injection of 50% glycerol (indicated as "0 h"), water deprivation should be started and be maintained at least 24 h. Free access to drinking water ad libitum is provided 12 h following glycerol injection until the readout systems are measured after 24 h or 48 h

References

1. Humes HD, Weinberg JM (1983) Alterations of renal tubular cell metabolism in acute renal failure. Miner Electrolyte Metab 9(4–6):290–305
2. Linkermann A, Skouta R, Himmerkus N, Mulay SR, Dewitz C, De Zen F, Prokai A, Zuchtriegel G, Krombach F, Welz PS, Weinlich R, Vanden Berghe T, Vandenabeele P, Pasparakis M, Bleich M, Weinberg JM, Reichel CA, Brasen JH, Kunzendorf U, Anders HJ, Stockwell BR, Green DR, Krautwald S (2014) Synchronized renal tubular cell death involves ferroptosis. Proc Natl Acad Sci U S A 111(47):16836–16841. https://doi.org/10.1073/pnas.1415518111
3. Linkermann A, Brasen JH, Himmerkus N, Liu S, Huber TB, Kunzendorf U, Krautwald S (2012) Rip1 (receptor-interacting protein kinase 1) mediates necroptosis and contributes to renal ischemia/reperfusion injury. Kidney Int 81(8):751–761. https://doi.org/10.1038/ki.2011.450
4. Linkermann A, Brasen JH, Darding M, Jin MK, Sanz AB, Heller JO, De Zen F, Weinlich R, Ortiz A, Walczak H, Weinberg JM, Green DR, Kunzendorf U, Krautwald S (2013) Two independent pathways of regulated necrosis mediate ischemia-reperfusion injury. Proc Natl Acad Sci U S A 110(29):12024–12029. https://doi.org/10.1073/pnas.1305538110
5. Singh P, Blantz RC, Rosenberger C, Gabbai FB, Schoeb TR, Thomson SC (2012) Aberrant tubuloglomerular feedback and HIF-1alpha confer resistance to ischemia after subtotal nephrectomy. J Am Soc Nephrol 23(3):483–493. https://doi.org/10.1681/asn.2011020130
6. Schrier RW (2002) Cancer therapy and renal injury. J Clin Invest 110(6):743–745. https://doi.org/10.1172/jci16568
7. Ries F, Klastersky J (1986) Nephrotoxicity induced by cancer chemotherapy with special emphasis on cisplatin toxicity. Am J Kidney Dis 8(5):368–379
8. Linkermann A, Himmerkus N, Rolver L, Keyser KA, Steen P, Brasen JH, Bleich M, Kunzendorf U, Krautwald S (2011) Renal tubular Fas ligand mediates fratricide in cisplatin-induced acute kidney failure. Kidney Int 79(2):169–178
9. Bolisetty S, Traylor A, Joseph R, Zarjou A, Agarwal A (2016) Proximal tubule-targeted heme oxygenase-1 in cisplatin-induced acute kidney injury. Am J Physiol Renal Physiol 310(5):F385–F394. https://doi.org/10.1152/ajprenal.00335.2015
10. Oken DE, Arce ML, Wilson DR (1966) Glycerol-induced hemoglobinuric acute renal failure in the rat. I. Micropuncture study of the development of oliguria. J Clin Invest 45(5):724–735. https://doi.org/10.1172/jci105387
11. Fahling M, Mathia S, Paliege A, Koesters R, Mrowka R, Peters H, Persson PB, Neumayer HH, Bachmann S, Rosenberger C (2013) Tubular von Hippel-Lindau knockout protects against rhabdomyolysis-induced AKI. J Am Soc Nephrol 24(11):1806–1819
12. Zager RA, Burkhart KM, Conrad DS, Gmur DJ (1995) Iron, heme oxygenase, and glutathione: effects on myohemoglobinuric proximal tubular injury. Kidney Int 48(5):1624–1634
13. Zarjou A, Bolisetty S, Joseph R, Traylor A, Apostolov EO, Arosio P, Balla J, Verlander J, Darshan D, Kuhn LC, Agarwal A (2013) Proximal tubule H-ferritin mediates iron trafficking in acute kidney injury. J Clin Invest 123(10):4423–4434

Chapter 14

Assessment of In Vivo Kidney Cell Death: Glomerular Injury

Wulf Tonnus, Moath Al-Mekhlafi, Florian Gembardt, Christian Hugo, and Andreas Linkermann

Abstract

The glomerulus functions as the filtration unit of the kidney. The mesangial, endothelial, and podocyte cells of the glomerulus exhibit the three clinically most important cell types, which are involved in diverse pathologic processes. Cell death has hardly been investigated in these cells but may be of critical importance to the pathogenesis of nephrotic syndrome, nephritic syndrome, focal segmental glomerulosclerosis (FSGS), mesangial proliferation, and thrombonic microangiopathy (which involves dysfunction and death of glomerular endothelial cells). The complexity of the glomerulus is frequently affected in autoimmune disorders, which may elicit cell death in mesangial cells and glomerular endothelia. Artificial antisera are used to induce anti-mesangial cell serum-induced mesangiolysis and selective endothelial cell injury, respectively. Genetic variations result in loss of function of podocytes and nephrotic syndrome, which may encompass similar cell death mechanisms as the ones that are observed in the model of secondary focal segmental glomerulosclerosis (FSGS). The following protocols describe our current arsenal to target glomerular cells in vivo.

Key words Anti-mesangial cell serum-induced mesangiolysis, Selective endothelial cell injury model, Secondary FSGS model

1 Anti-Mesangial Cell Serum-Induced Mesangiolysis

1.1 Introduction

Mesangial cells (MCs) are critical players in initiation and progression of glomerular disease which most commonly presents as mesangioproliferation. Mesangial proliferative disease, as prominently described in IgA nephropathy, is the most frequent glomerulonephritis in man.

Sheep antisera raised against murine mesangial cells induced acute glomerulopathy and MC injury [1, 2]. The anti-mouse mesangial cell serum (MC-Serum)-treated mice exhibit a duplication of glomerular basement membranes, mesangiolysis, subendothelial and mesangial electron-dense deposits and foot process effacement [1]. For aggravation of mesangiolysis, this anti-MC model has been

Wulf Tonnus and Moath Al-Mekhlafi contributed equally to this work.

Adrian T. Ting (ed.), *Programmed Necrosis: Methods and Protocols*, Methods in Molecular Biology, vol. 1857, https://doi.org/10.1007/978-1-4939-8754-2_14, © Springer Science+Business Media, LLC, part of Springer Nature 2018

modified by application of LPS before the first injection of anti-MC antibody [2]. Until today, it is entirely elusive which mode of programmed cell death causes mesangiolysis. On day 2–3 after initiation, the disease model reveals pronounced mesangiolysis, although to a varying extent in individual glomerulus. Consecutively, regeneration processes are initiated that include cellular repopulation up to days 6–13 [2]. Whether cell death signals are involved in the initiation of a regenerative signal has not been investigated.

1.2 Materials

1.2.1 Cell Culture

1. DMEM/F12 media supplemented with 10% heat-inactivated fetal calf serum (FCS) and 100 U penicillin–0.1 mg/mL streptomycin (1% P/S).
2. 175 cm^2 culture flask.
3. Phosphate-buffered saline (PBS).
4. 1× trypsin–EDTA
5. Cell scraper.
6. Dry centrifuge.

1.2.2 Other Materials

1. 27-gauge microcannula.
2. 31-gauge microcannula.
3. 1 mL syringe.

1.3 Method

The anti-mouse mesangial cell (MC) serum is prepared by immunizing sheep with SV40-transformed mouse mesangial MES 13 cells.

1.3.1 Cell Culture

1. 4×10^5 SV40 MES 13 cells in a final volume of 15 mL. Allow 80–90% confluence and harvest using trypsin–EDTA.
2. Wash 3 times with medium and spin the cell suspension at 350 × *g* at 4 °C for 5 min.
3. Freeze dry and weigh the pellet.
4. Resuspend the pellet in PBS (1 mL PBS per mg pellet) and transfer to 1.5 mL Eppendorf tube. Aliquots of 300 μL portions may be helpful to save the same cellular preparation for several immunizations—avoid frequent freeze–thaw cycles. Centrifuge all aliquots at 1300 × *g* at 4 °C for 5 min, and freeze the pellet at −80 °C until use.

1.3.2 Sheep Immunization

1. Inject the sheep subcutaneously with the mixture of 300 μL of SV40 MES 13 antigen in Freund's complete adjuvant.
2. 3 weeks later, inject the sheep s.c. with another 300 μL of SV40 MES 13 antigen in Freund's incomplete adjuvant (first boosting).
3. The second and third boosting followed 3 and 6 weeks after the first boosting subsequently.
4. 6 months after immunizing, the sheep undergoes plasmapheresis. The plasma will clot after adding calcium, and serum can be obtained by centrifugation.
5. Decomplement sera by heating at 56 °C for 30 min and prepare aliquots for storage at −80 °C.

1.3.3 Induction of Mesangiolysis

1. To aggravate mesangial injury, inject the mice i.p, using 27-gauge microcannula, with 1 mg/kg LPS 2 h before applying the mesangial cell (MC) serum.
2. Two and 26 h following LPS application, mice are injected 5 mL/kg body weight of MC serum i.v. via the tail vein.
3. As a control, inject the mice with sheep preimmune control serum.
4. On day 2–3 the disease model will reveal overall pronounced mesangiolysis at varying degrees in individual glomerulus.
5. On day 10–13, cellular repopulation reaches a maximum after injury [2].

2 Glomerular Endothelial Cell Injury Model

2.1 Introduction

The renal endothelium represents the major nutritional and functional surface between the blood and parenchymal renal cells. They exhibit a specialized microvascular cell type involved in the regulation of glomerular ultrafiltration [3]. Endothelial cell injury due to a severe glomerular lesion can result in sclerosis at the injured site, which is inevitably followed by nephron loss and progressive renal dysfunction [4]. Lectin concanavalin A (conA) is a widely used model for glomerular endothelial cell injury. The model was first described by Golbus and Wilson [5] and allows targeting by a specific anti-concanavalin A antibody. This model is characterized by endothelial cell death, the signaling pathway of which remains elusive, and is pathophysiologically accompanied by subendothelial immune deposit, complement activation, and infiltration by platelets and neutrophils [6], nicely supporting the recently published concept of necroinflammation [7, 8].

2.2 Materials

1. A vaporizer for inhalative anesthetics (isoflurane).
2. An autoregulated heating plate (36 °C–37 °C)
3. Surgical instruments: a needle-holder, a precision scissors, anatomic forceps, pointed forceps, larger forceps, 6-0 sutures, 31-gauge microcannula (length 40 mm).
4. A surgical microscope.
5. A set of two microaneurysm clamps.
6. A 10–20 mL syringe containing prewarmed PBS (37 °C).
7. An analgesic with favorable pharmacokinetics, e.g., buprenorphine–HCl.

2.3 Method

2.3.1 Anti-Concanavalin A Serum

The rabbit anti-Con A antiserum is prepared by immunizing rabbits at monthly intervals with 3 mg of Con A in Freund's incomplete adjuvant [5]. See the generation of the anti-mesangial serum for details (*see* Subheading 1.3.2).

2.3.2 Renal Arterial Perfusion in Mice

1. Use at least 8 male mice per group at the age of 12–14 weeks (25–30 g).
2. Anesthetize mice using isoflurane.
3. Upon narcosis, inject buprenorphine–HCl s.c. or i.m.
4. A paravertebral incision at the left flank (15 mm) is set using a precision scissors.
5. Mobilize the kidney and carefully remove the surrounding fat pads.
6. Upon microscope control, the left renal artery is cannulated via the abdominal aorta with 31-gauge microcannula. The cannula is connected to a 20-gauge cannula that shall be placed i.v. (*see* **Note 1**).
7. The kidney is perfused with 150 μL PBS to remove blood cells.
8. 200 μg of ConA followed by 150 μL PBS is injected.
9. 18 mg of rabbit anti-ConA antiserum is applied followed by another 150 μL PBS [9, 10].
10. The abdominal aorta is clamped for vessel closure using 1.5 μL tissue glue before removal of the clamp.
11. The wound is closed with continuous sutures and mice are placed under a heat lamp for 1 h to allow recovery.
12. 48 h later, standard readout systems may be evaluated.

2.4 Notes

1. The complete device contains a loading volume of 150 μL and allows for consecutive injections of PBS, ConA, and anti-ConA by an assisting person with minimal movement of the microcannula after insertion into the renal artery [9, 10].

3 Targeting Podocytes—The Secondary FSGS Model

3.1 Introduction

Podocytes are highly specialized cells of the kidney glomerulus and are critical for glomerular function. Podocyte function together with mesangial cells [11]. Podocytes are unable to self-renew and therefore their replacement depends on progenitor cells, such as partial epithelial cells and cells of renin lineage (CoRL) [12–14]. Podocyte damage and/or podocyte death cause foot process effacement, proteinuria and focal segmental glomerulosclerosis (FSGS) [15, 16]. FSGS is leading cause of nephrotic syndrome in adult patients, and it is a major task for cell death and podocyte researchers to clarify the involvement of regulated cell death in podocyte loss, which may exhibit a druggable disorder. In addition, loss of podocytes is a critical determinant of glomerulosclerosis [17–19].

3.2 Method

3.2.1 Sheep Anti-Rabbit Glomeruli Antisera

1. Sheep anti-rabbit glomeruli sera (often mistaken in the literature as an "anti-rabbit glomeruli antibody") is produced in an immunized sheep 3–4 times at 2 week intervals with lyophilized whole rabbit glomeruli solubilized in Freund's complete adjuvant at the first immunization, and subsequently in Freund's incomplete adjuvant at later immunizations [20].
2. Rabbit glomeruli for immunization were published to be isolated by differential sieving techniques and should contain less than one tubular fraction per 100 glomeruli [20, 21]. However, to the best of our knowledge, the exact protocol has never been published and it was never used by us.
3. For in vivo experiments, sera should be decomplemented by heating at 56 °C for 30 min.

3.2.2 Experimental FSGS

1. Adult male mice are preferentially used.
2. Two doses of sheep anti-glomerular antibody are injected i.p. at 12 mg/20 g body weight (*see* **Note 1** of Subheading 3.3).
3. Upon progression to the end of the variable incubation period, mice are routinely harvested (*see* **Note 2** of Subheading 3.3) for histopathological and immunofluorescence readouts.

3.3 Notes

1. 24 h is enough to see abrupt podocyte depletion accompanied by glomerulosclerosis [22, 23].
2. On day 3–4 of disease, the podocytes number can be decreased by 30%–40% [14].

4 Concluding Remarks

Two virtually separate fields of research benefit from the investigation of the mentioned protocols. It is recognized that other models may be suited for the investigation of cell death in the kidney as well, such as the folic acid-induced AKI model [24], the nephrotoxic serum nephritis [25], models of ANCA-vasculitis [26], snake venom induced mesangiolysis [27], and countless others. We also used models of contrast-induced AKI (CIAKI), but cell death did not contribute to the pathophysiology of these models in our hands [27]. However, we chose to present the ones mentioned in this and the previous chapters because of our own experience with these models and their relatively easy setup. Whereas cell death researchers find appropriate in vivo models for their analysis of necroptosis, ferroptosis, MPT-RN, and probably several other cell death pathways, translational kidney scientists use inhibitors of cell death to approach these models from a therapeutical perspective. The kidney, like no other organ, therefore represents a beautiful example of interdisciplinary research with a wealth of unstudied opportunities.

References

1. Yo Y, Braun MC, Barisoni L, Mobaraki H, Lu H, Shrivastav S, Owens J, Kopp JB (2003) Anti-mouse mesangial cell serum induces acute glomerulonephropathy in mice. Nephron Exp Nephrol 93(3):e92–e106. https://doi.org/10.1159/000069551
2. Starke C, Betz H, Hickmann L, Lachmann P, Neubauer B, Kopp JB, Sequeira-Lopez ML, Gomez RA, Hohenstein B, Todorov VT, Hugo CP (2015) Renin lineage cells repopulate the glomerular mesangium after injury. J Am Soc Nephrol 26(1):48–54. https://doi.org/10.1681/ASN.2014030265
3. Nangaku M, Shankland SJ, Couser WG, Johnson RJ (1998) A new model of renal microvascular injury. Curr Opin Nephrol Hy 7(4):457–462. https://doi.org/10.1097/00041552-199807000-00018
4. Lee LK, Meyer TW, Pollock AS, Lovett DH (1995) Endothelial-cell injury initiates glomerular sclerosis in the rat remnant kidney. J Clin Invest 96(2):953–964. https://doi.org/10.1172/Jci118143
5. Golbus SM, Wilson CB (1979) Experimental glomerulonephritis induced by in situ formation of immune-complexes in glomerular Capillary Wall. Kidney Int 16(2):148–157. https://doi.org/10.1038/Ki.1979.116
6. Johnson RJ, Garcia RL, Pritzl P, Alpers CE (1990) Platelets mediate glomerular cell proliferation in immune complex nephritis induced by anti-mesangial cell antibodies in the rat. Am J Pathol 136(2):369–374
7. Land WG, Agostinis P, Gasser S, Garg AD, Linkermann A (2016) Transplantation and damage-associated molecular patterns (DAMPs). Am J Transplant 16(12):3338–3361
8. Land WG, Agostinis P, Gasser S, Garg AD, Linkermann A (2016) DAMP - induced allograft and tumor rejection: the circle is closing. Am J Transplant. https://doi.org/10.1111/ajt.14012
9. Hohenstein B, Braun A, Amann KU, Johnson RJ, Hugo CPM (2008) A murine model of site-specific renal microvascular endothelial injury and thrombotic microangiopathy. Nephrol Dial Transpl 23(4):1144–1156. https://doi.org/10.1093/ndt/gfm774
10. Sradnick J, Rong S, Luedemann A, Parmentier SP, Bartaun C, Todorov VT, Gueler F, Hugo CP, Hohenstein B (2016) Extrarenal progenitor cells do not contribute to renal endothelial repair. J Am Soc Nephrol 27(6):1714–1726. https://doi.org/10.1681/Asn.2015030321
11. Reiser J, Altintas MM (2016) Podocytes. F1000Res 5:F1000 Faculty Rev-1114. https://doi.org/10.12688/f1000research.7255.1
12. Ohse T, Vaughan MR, Kopp JB, Krofft RD, Marshall CB, Chang AM, Hudkins KL, Alpers CE, Pippin JW, Shankland SJ (2010) De novo expression of podocyte proteins in parietal epithelial cells during experimental glomerular disease. Am J Physiol-Renal 298(3):F702–F711. https://doi.org/10.1152/ajprenal.00428.2009
13. Lazzeri E, Romagnani P (2015) Differentiation of parietal epithelial cells into podocytes. Nat Rev Nephrol 11(1):7–U8888. https://doi.org/10.1038/nrneph.2014.218
14. Lichtnekert J, Kaverina NV, Eng DG, Gross KW, Kutz JN, Pippin JW, Shankland SJ (2016) Renin-angiotensin-aldosterone system inhibition increases Podocyte derivation from cells of renin lineage. J Am Soc Nephrol 27(12):3611–3627. https://doi.org/10.1681/Asn.2015080877
15. Kim JS, Han BG, Choi SO, Cha SK (2016) Secondary focal segmental Glomerulosclerosis: from Podocyte injury to Glomerulosclerosis. Biomed Res Int. https://doi.org/10.1155/2016/1630365
16. D'Agati VD, Kaskel FJ, Falk RJ (2011) Focal segmental glomerulosclerosis. N Engl J Med 365(25):2398–2411. https://doi.org/10.1056/NEJMra1106556
17. Fogo AB (2007) Mechanisms of progression of chronic kidney disease. Pediatr Nephrol 22(12):2011–2022. https://doi.org/10.1007/s00467-007-0524-0
18. Shankland SJ (2006) The podocyte's response to injury: role in proteinuria and glomerulosclerosis. Kidney Int 69(12):2131–2147. https://doi.org/10.1038/sj.ki.5000410
19. Wiggins RC (2007) The spectrum of podocytopathies: a unifying view of glomerular diseases. Kidney Int 71(12):1205–1214. https://doi.org/10.1038/sj.ki.5002222
20. Ophascharoensuk V, Pippin JW, Gordon KL, Shankland SJ, Couser WG, Johnson RJ (1998) Role of intrinsic renal cells versus infiltrating cells in glomerular crescent formation. Kidney Int 54(2):416–425. https://doi.org/10.1046/j.1523-1755.1998.00003.x
21. Couser WG, Darby C, Salant DJ, Adler S, Stilmant MM, Lowenstein LM (1985) Anti-GBM antibody-induced proteinuria in isolated perfused rat kidney. Am J Phys 249(2 Pt 2):F241–F250

22. Pippin JW, Glenn ST, Krofft RD, Rusiniak ME, Alpers CE, Hudkins K, Duffield JS, Gross KW, Shankland SJ (2014) Cells of renin lineage take on a podocyte phenotype in aging nephropathy. Am J Physiol Renal Physiol 306(10):F1198–F1209. https://doi.org/10.1152/ajprenal.00699.2013
23. Zhang J, Yanez D, Floege A, Lichtnekert J, Krofft RD, Liu ZH, Pippin JW, Shankland SJ (2015) ACE-inhibition increases podocyte number in experimental glomerular disease independent of proliferation. J Renin Angiotensin Aldosterone Syst 16(2):234–248. https://doi.org/10.1177/1470320314543910
24. Martin-Sanchez D, Ruiz-Andres O, Poveda J, Carrasco S, Cannata-Ortiz P, Sanchez-Nino MD, Ruiz OM, Egido J, Linkermann A, Ortiz A, Sanz AB (2017) Ferroptosis, but not Necroptosis, is important in nephrotoxic folic acid-induced AKI. J Am Soc Nephrol 28(1):218–229. https://doi.org/10.1681/ASN.2015121376
25. Turner JE, Paust HJ, Steinmetz OM, Peters A, Meyer-Schwesinger C, Heymann F, Helmchen U, Fehr S, Horuk R, Wenzel U, Kurts C, Mittrucker HW, Stahl RA, Panzer U (2008) CCR5 deficiency aggravates crescentic glomerulonephritis in mice. J Immunol 181(9):6546–6556
26. Schreiber A, Kettritz R (2013) The neutrophil in antineutrophil cytoplasmic autoantibody-associated vasculitis. J Leukoc Biol 94(4):623–631. https://doi.org/10.1189/jlb.1012525
27. Chen L, Lu Y, Wen J, Wang X, Wu L, Wu D, Sun X, Fu B, Yin Z, Jiang H, Chen X (2016) Comparative proteomics analysis of mouse Habu nephritis models with and without unilateral nephrectomy. Cellular physiology and biochemistry: international journal of experimental cellular physiology. Biochem Pharmacol 39(5):1761–1776. https://doi.org/10.1159/000447876

Chapter 15

Detection of Necroptosis by Phospho-RIPK3 Immunohistochemical Labeling

Joshua D. Webster, Margaret Solon, Susan Haller, and Kim Newton

Abstract

Activation of the kinase RIPK3 (receptor interacting protein kinase 3) is a hallmark of cells dying by necroptosis. RIPK3 phosphorylates both itself and the pseudokinase MLKL (mixed lineage kinase-like) resulting in MLKL translocation to membranes and cell lysis. Antibodies recognizing RIPK3 autophosphorylation or the RIPK3-dependent phosphorylation sites on MLKL have therefore been used to monitor necroptosis induction. Here we describe immunohistochemical labeling for autophosphorylated mouse RIPK3 as a means of detecting cells undergoing necroptosis in mouse tissues.

Key words Necroptosis, RIPK3, Autophosphorylation, Immunohistochemistry, Mouse

1 Introduction

The contribution of RIPK3/MLKL-dependent necroptosis to various pathologies has not been explored extensively because of a paucity of quality antibodies for staining necroptotic cells in tissue sections. Many studies have assessed total RIPK3 or MLKL levels in tissues, but increased RIPK3 or MLKL abundance does not necessarily indicate pathway activation. Better indicators of necroptosis signaling are RIPK3 autophosphorylation and RIPK3-mediated phosphorylation of MLKL. Autophosphorylation of human RIPK3 on Ser^{227} [1] or mouse RIPK3 on Thr^{231} and Ser^{232} [2] is required for recruitment of MLKL, whereas subsequent phosphorylation of human MLKL on Thr^{357} and Ser^{358} [1] or mouse MLKL on Ser^{345} and Ser^{347} [3, 4] is required for membrane disruption. An antibody recognizing human MLKL phosphorylated on Thr^{357} and Ser^{358} has been shown to immunolabel cells in liver biopsies from patients with drug-induced liver injury [5] and in kidney biopsies with evidence of oxalate crystal-related acute kidney injury [6]. However, an antibody that detects phosphorylated mouse MLKL in tissue sections has yet to be reported for use in preclinical disease models. Therefore, a role for necroptosis has largely been determined by

Adrian T. Ting (ed.), *Programmed Necrosis: Methods and Protocols*, Methods in Molecular Biology, vol. 1857,
https://doi.org/10.1007/978-1-4939-8754-2_15,

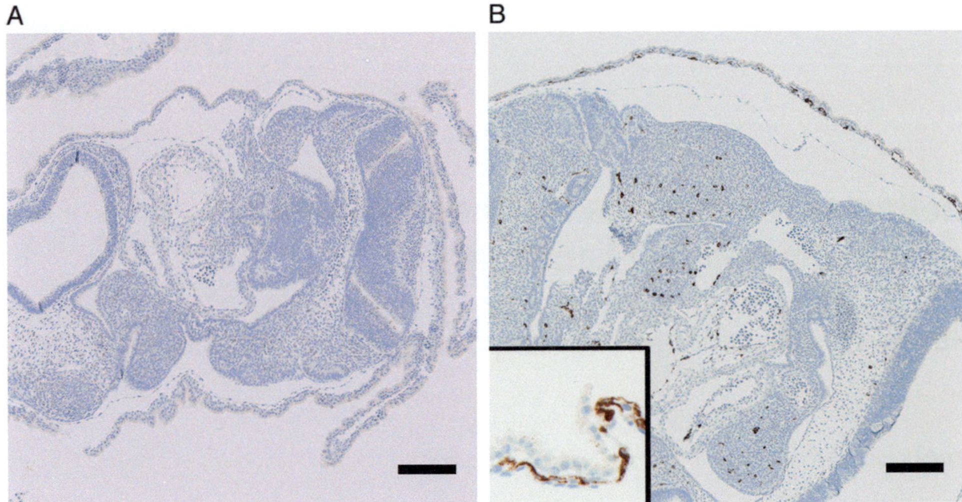

Fig. 1 Immunohistochemical labeling of phospho-RIPK3 $Thr^{231}Ser^{232}$ in E10.5 mouse embryos. Wild-type (**a**) and *Casp8*$^{-/-}$ (**b**) mouse embryos immunolabeled with rabbit anti-mouse phospho-RIPK3 $Thr^{231}Ser^{232}$ antibody. *Casp8*$^{-/-}$ cells containing phospho-RIPK3 labeled brown. Insert shows endothelial cell labeling in the yolk sac. Labeling was cytoplasmic and multifocal. Scale bar, 200 μm

genetic rescue experiments with MLKL-deficient mice. Recently, we identified an anti-mouse phospho-RIPK3 Thr^{231}, Ser^{232} monoclonal antibody that is suitable for western blotting, immunoprecipitation, and immunohistochemical detection of necroptosis [7]. Here we describe the protocol for immunohistochemical identification of necroptotic cells in formalin-fixed, paraffin-embedded mouse tissue sections. As an example, immunolabeling of caspase-8-deficient mouse embryos [8] at embryonic day 10.5 (E10.5), which die due to RIPK3/MLKL-dependent necroptosis [9–11], is shown in Fig. 1.

2 Materials

2.1 Tissue Fixation and Sectioning

1. Tissue sponges (Bioindustrial Products) (*see* **Note 1**).
2. Tissue cassettes (Thermo Scientific) (*see* **Note 2**).
3. 10% neutral buffered formalin fixative (handle in fume cupboard with appropriate personal protective equipment).
4. Tissue Tek VIP 5 Tissue Processor (Sakura).
5. 70% v/v ethanol in water.
6. Flex 95 and Flex 100 dehydrant solutions (Richard Allan Scientific).
7. Xylene.
8. Paraplast Tissue Embedding Medium (McCormick Scientific).
9. HistoStar Embedding Workstation (Thermo Scientific).

10. Embedding base molds (Sakura).
11. RM2255 Microtome (Leica).
12. MX35 Ultra Microtome Blades (Thermo Scientific).
13. Superfrost Plus slides (Thermo Scientific).
14. Tissue Flotation Bath set at 46 °C.
15. Slide-drying hot plate set at 57 °C.
16. Lab oven set at 60 °C.

2.2 Immunohistochemistry

1. Discovery XT automated slide staining system (Ventana).
2. Cell conditioning solution CC1 (Ventana).
3. Discovery Goat Ig Block (Ventana).
4. Discovery Amp HQ kit (Ventana).
5. Discovery anti-Rabbit HQ (Ventana).
6. Discovery anti-HQ HRP (Ventana).
7. Discovery Amplification anti-HQ HRP multimer (Ventana).
8. Discovery ChromoMap DAB kit (Ventana).
9. Hematoxylin II (Ventana).
10. Bluing reagent (Ventana).
11. GEN135-35-9 rabbit anti-mouse phospho-RIPK3 $Thr^{231}Ser^{232}$ monoclonal antibody (Genentech).
12. Phosphate buffered saline (PBS): 8 g NaCl, 0.2 g KCl, 1.13 g Na_2HPO_4, 0.2 g KH_2PO_4, add water to 1 L and adjust pH to 7.2 with 6 N HCl.
13. 3% w/v bovine serum albumin (BSA) in PBS (*see* **Note 3**).
14. Control tissues or cell pellets (*see* **Note 4**).
15. Dawn Dishfoam.
16. Staining buckets and 20-slide holder (Sakura).
17. 100% and 95% v/v ethanol in water.
18. Xylene (handle in fume cupboard with appropriate personal protective equipment).
19. Mounting medium.
20. Glass coverslips.
21. Tissue Tek Glas g2 Coverslipper (Sakura).

3 Methods

3.1 Tissue Fixation and Sectioning

1. Place tissue sample onto a foam sponge (*see* **Note 1**) inside a labeled tissue cassette (*see* **Note 2**).
2. Place the cassettes containing tissues in a jar of 10% neutral buffered formalin to fix at room temperature for 24 h (*see* **Notes 5–7**).

3. Transfer tissue cassettes to the automated tissue processor for dehydration: incubate in 70% v/v ethanol in water for 1 h, then twice in Flex 95 for 1 h, then three times in Flex 100 for 1 h, and then three times in xylene for 1 h. Perform all incubations at room temperature (*see* **Note 8**).
4. Infiltrate tissues in processor with paraplast tissue embedding medium at 60 °C for 1 h. Repeat incubation with fresh embedding medium, and then again with a third change of embedding medium.
5. Select an appropriately sized embedding mold (*see* **Note 9**) and add molten embedding medium from the embedding workstation so that the mold is approximately a quarter full.
6. Use warmed forceps to transfer the infiltrated tissue into the mold, and then place the mold on the 5 °C cold plate of the embedding workstation. Use forceps and gentle pressure to keep the tissue in the desired orientation as the paraffin cools and anchors the tissue in place.
7. Place the labeled tissue cassette lid over the top of the mold and dispense sufficient embedding medium over the lid to fill the mold. Transfer the mold and its lid to the 5 °C cold plate for 10 min or until the paraffin has hardened enough for manual release of the tissue block from the base mold.
8. Trim blocks at room temperature on a RM2255 microtome: cut 4–10 μm at a time in continuous motorized mode with rotary knob sectioning speed set to seven until all surfaces of tissue in a block are visible.
9. Soak trimmed blocks face down in icy water for 10 min and then manually section 4 μm thick tissue ribbons using disposable microtome blades (*see* **Note 10**).
10. Float tissue ribbons in a 46 °C tissue flotation bath for 30 s and then lift individual tissue sections onto Superfrost Plus slides.
11. Place slides on a 57 °C hot plate for 30 min to overnight until sample is free of visible moisture.
12. Dry slides in a 60 °C oven for 20 min and proceed with immunohistochemical labeling.

3.2 Immunohistochemistry

1. Perform staining on the Ventana Discovery XT using the procedure template called Res IHC Discovery HQ-HRP. Parameter selection is shown in Table 1.
2. Print slide labels and attach to unbaked slides.
3. Place slides on the Ventana Discovery XT carousel after refilling all bulk reagent bottles.
4. Initiate the IHC run, which is summarized as follows:

(a) Deparaffinize.

(b) Perform antigen retrieval with cell conditioning solution CC1 for 48 min at 100 °C.

(c) Block nonspecific binding sites with goat Ig block for 4 min.

(d) Stain with 5 μg/mL GEN135-35-9 rabbit anti-mouse phospho-RIPK3 $Thr^{231}Ser^{232}$ monoclonal antibody in 3% BSA/PBS for 1 h at 37°C (*see* **Note 11**).

(e) Detect bound GEN135-35-9 by incubating slides with anti-rabbit HQ for 16 min, followed by anti-HQ HRP for 16 min.

(f) Amplify the signal with TSA HQ and H_2O_2 from the Discovery Amp HQ kit for 12 min, followed by anti-HQ HRP multimer for 8 min.

(g) Develop the slides with DAB (3,3′-diaminobenzidine) for 8 min.

(h) Counterstain with Hematoxylin II for 4 min followed by Bluing reagent for 4 min.

5. Once the run is completed, remove the slides from the machine carousel and place them in a 20-slide holder.
6. Dip the slides in and out of soapy water (*see* **Note 12**) for 1 min to remove the liquid coverslips.
7. Rinse slides briefly in water to remove the soap.
8. Dehydrate slides by incubating twice in 95% v/v ethanol in water for 1 min, twice in 100% ethanol for 1 min, and then twice in xylene for 1 min.
9. Coverslip slides with permanent mounting medium using the Tissue Tek Glas g2 Coverslipper.

4 Notes

1. Uni-cassettes 30 mm × 26 mm × 5 mm deep are suitable for most routine samples, but larger samples may require cassettes that are 10 mm deep. Tissues should not be compressed when the cassette lid is closed.
2. A foam insert in the tissue cassette helps hold the tissue in place during processing but is not always necessary. Delicate samples such as E10.5 mouse embryos can be placed in a Cell Safe Mesh Biopsy Cassette (Cancer Diagnostics, Inc), which fits inside a standard Uni-cassette to limit tissue loss or damage during automated processing.
3. The 3% w/v BSA in PBS antibody diluent should be filter-sterilized with a 0.2 μm filter and stored at 4 °C when not in use.

Table 1
Set-up parameters for labeling mouse phospho-RIPK3 with the Ventana Res IHC DISCOVERY HQ-HRP procedure

1	Tissue sample [selected]
2	Paraffin [selected]
3	Deparaffinization [selected]
4	Cell conditioning [selected]
5	CC1 [selected]
6	CC1 8 min [selected]
7	CC1 16 min [selected]
8	CC1 24 min [selected]
9	CC1 32 min [selected]
10	CC1 40 min [selected]
11	CC1 48 min [selected]
12	Disable discovery inhibitor [selected]
13	Antibody [selected]
14	Antibody titration [selected]
15	Standard titration Ab Inc. [selected]
16	Antibody blocking [selected]
17	Apply one drop of [blocker 1] (antibody blocking), apply coverslip, and incubate for 4 min
18	Incubate for [60 min] (primary antibody)
19	Apply one drop of [Anti-Rabbit HQ] (Detection #1), apply coverslip, and incubate for [16 min]
20	Apply coverslip, one drop of [Anti-HQ HRP] (Conjugate #1), and incubate for [16 min]
21	Discovery amplification HQ [selected]
22	Apply one drop of DISC AMP TSA HQ and one drop of DISC AMP H2O2 HQ, apply coverslip, incubate for [12 min]
23	Apply one drop of [DISC anti-HQ HRP] (DISCOVERY AMP Multimer), and incubate for [8 min]
24	DAB [selected]
25	Counterstain [selected]
26	Standard [selected]
27	Apply one drop of [Hematoxylin II] (Counterstain), apply coverslip, and incubate for [4 min]
28	Post counterstain [selected]
29	Apply one drop of [BLUING REAGENT] (Post Counterstain), apply coverslip, and incubate for [4 min]

4. Mouse L-929 cells (ATCC CCL-1) that are untreated or stimulated for 4 h with 100 ng/mL murine TNF (tumor necrosis factor) and 20 μM pan-caspase inhibitor Z-VAD-FMK (carbobenzoxy-valyl-alanyl-aspartyl-[O-methyl]-fluoromethylketone) can be used as negative and positive controls for phospho-RIPK3 $Thr^{231}Ser^{232}$ staining, respectively. Cells from approximately 10 × 15 cm dishes should be pelleted, fixed in 10% neutral buffered formalin for 24 h and then embedded in paraffin for sectioning. The embedding process for cell pellets is the same as for tissues except that 40 min rather than 1 h processing times are used.
5. Ideally tissue samples should be no greater than 3 mm in thickness to achieve adequate fixation.
6. Use at least 15 tissue volumes of 10% neutral buffered formalin to achieve adequate fixation.
7. Fixing tissues for longer than 24 h can compromise antigen availability and should be avoided. Fixed tissues can be stored in 70% v/v ethanol in water at room temperature prior to further processing. Alternatively, fixed mouse embryos (from E10.5 through 13.5) can be stored in PBS at 4 °C for 1–2 weeks before transfer to 70% v/v ethanol in water.
8. An alternating vacuum and pressure setting and a slow mix cycle are recommended for routine tissue samples during tissue dehydration and infiltration (described in Subheading 3.1, **steps 3** and **4**). However, a vacuum setting is not recommended for delicate tissues such as E10.5 mouse embryos. Processing times for delicate tissues should be reduced from 1 h to 20 min.
9. Base molds should be selected such that at least 3 mm separates the edges of tissue from the edges of the mold.
10. To prevent scratches on tissue sections, move the microtome blade in the blade holder so that an unused portion of the blade is selected for cutting every ten blocks. Blades should be replaced more frequently if tissues present surfaces of varying hardness, as in large mouse embryos or in blocks that contain mixed tissues.
11. Monoclonal rabbit IgG can be used as an isotype control.
12. Add two pumps of Dawn soap to a 300 mL staining bucket then fill with water.

References

1. Sun L, Wang H, Wang Z, He S, Chen S, Liao D, Wang L, Yan J, Liu W, Lei X, Wang X (2012) Mixed lineage kinase domain-like protein mediates necrosis signaling downstream of RIP3 kinase. Cell 148(1–2):213–227. https://doi.org/10.1016/j.cell.2011.11.031
2. Chen W, Zhou Z, Li L, Zhong CQ, Zheng X, Wu X, Zhang Y, Ma H, Huang D, Li W, Xia Z,

Han J (2013) Diverse sequence determinants control human and mouse receptor interacting protein 3 (RIP3) and mixed lineage kinase domain-like (MLKL) interaction in necroptotic signaling. J Biol Chem 288(23):16247–16261. https://doi.org/10.1074/jbc.M112.435545

3. Murphy JM, Czabotar PE, Hildebrand JM, Lucet IS, Zhang JG, Alvarez-Diaz S, Lewis R, Lalaoui N, Metcalf D, Webb AI, Young SN, Varghese LN, Tannahill GM, Hatchell EC, Majewski IJ, Okamoto T, Dobson RC, Hilton DJ, Babon JJ, Nicola NA, Strasser A, Silke J, Alexander WS (2013) The pseudokinase MLKL mediates necroptosis via a molecular switch mechanism. Immunity 39(3):443–453. https://doi.org/10.1016/j.immuni.2013.06.018
4. Rodriguez DA, Weinlich R, Brown S, Guy C, Fitzgerald P, Dillon CP, Oberst A, Quarato G, Low J, Cripps JG, Chen T, Green DR (2016) Characterization of RIPK3-mediated phosphorylation of the activation loop of MLKL during necroptosis. Cell Death Differ 23(1):76–88. https://doi.org/10.1038/cdd.2015.70
5. Wang H, Sun L, Su L, Rizo J, Liu L, Wang LF, Wang FS, Wang X (2014) Mixed lineage kinase domain-like protein MLKL causes necrotic membrane disruption upon phosphorylation by RIP3. Mol Cell 54(1):133–146. https://doi.org/10.1016/j.molcel.2014.03.003
6. Mulay SR, Desai J, Kumar SV, Eberhard JN, Thomasova D, Romoli S, Grigorescu M, Kulkarni OP, Popper B, Vielhauer V, Zuchtriegel G, Reichel C, Brasen JH, Romagnani P, Bilyy R, Munoz LE, Herrmann M, Liapis H, Krautwald S, Linkermann A, Anders HJ (2016) Cytotoxicity of crystals involves RIPK3-MLKL-mediated necroptosis. Nat Commun 7:10274. https://doi.org/10.1038/ncomms10274
7. Newton K, Wickliffe KE, Maltzman A, Dugger DL, Strasser A, Pham VC, Lill JR, Roose-Girma M, Warming S, Solon M, Ngu H, Webster JD, Dixit VM (2016) RIPK1 inhibits ZBP1-driven necroptosis during development. Nature 540(7631):129–133. https://doi.org/10.1038/nature20559
8. Newton K, Dugger DL, Wickliffe KE, Kapoor N, de Almagro MC, Vucic D, Komuves L, Ferrando RE, French DM, Webster J, Roose-Girma M, Warming S, Dixit VM (2014) Activity of protein kinase RIPK3 determines whether cells die by necroptosis or apoptosis. Science 343(6177):1357–1360. https://doi.org/10.1126/science.1249361
9. Kaiser WJ, Upton JW, Long AB, Livingston-Rosanoff D, Daley-Bauer LP, Hakem R, Caspary T, Mocarski ES (2011) RIP3 mediates the embryonic lethality of caspase-8-deficient mice. Nature 471(7338):368–372. https://doi.org/10.1038/nature09857
10. Oberst A, Dillon CP, Weinlich R, McCormick LL, Fitzgerald P, Pop C, Hakem R, Salvesen GS, Green DR (2011) Catalytic activity of the caspase-8-FLIP(L) complex inhibits RIPK3-dependent necrosis. Nature 471(7338):363–367. https://doi.org/10.1038/nature09852
11. Alvarez-Diaz S, Dillon CP, Lalaoui N, Tanzer MC, Rodriguez DA, Lin A, Lebois M, Hakem R, Josefsson EC, O'Reilly LA, Silke J, Alexander WS, Green DR, Strasser A (2016) The Pseudokinase MLKL and the kinase RIPK3 have distinct roles in autoimmune disease caused by loss of death-receptor-induced apoptosis. Immunity 45(3):513–526. https://doi.org/10.1016/j.immuni.2016.07.016

Chapter 16

Characterization of the TNFR1-SC Using "Modified Tandem Affinity Purification" in Conjunction with Liquid Chromatography–Mass Spectrometry (LC-MS)

Matthias Reichert, Amandeep Bhamra, Sebastian Kupka, and Henning Walczak

Abstract

Mass spectrometry enables the unbiased characterization of protein complexes. The success of this approach and the amount of information that can be retrieved are highly dependent on the achieved purity of the protein complex to be analyzed. Here we describe a modified tandem affinity purification (moTAP) approach which can be used to isolate the tumor necrosis factor receptor 1 signaling complex for subsequent analysis by liquid chromatography–tandem mass spectrometry. Indeed, this approach can easily be adapted to the isolation of other membrane-bound and intracellular signaling complexes. This methodology allows for a highly sensitive analysis and characterization of complex components, including posttranslational modifications and for the identification of novel complex components.

Key words Tandem affinity purification, Immunoprecipitation, Proteomics, Tumor necrosis factor (TNF), TNF receptor 1 signaling complex (TNFR1-SC), Liquid chromatography-tandem mass spectrometry (LC-MS/MS)

1 Introduction

Liquid chromatography-tandem mass spectrometry (LC-MS/MS) is a powerful methodology for the unbiased analysis of complex protein assemblies [1]. This method enables researchers to identify unknown components of protein complexes of interest as well as posttranslational modifications thereof. In the protein characterization approach referred to as "bottom-up," the proteins are enzymatically digested by proteases resulting in a mixture of short peptides which are then separated by reverse-phase liquid chroma-

Amandeep Bhamra and Sebastian Kupka contributed equally to this work.

Adrian T. Ting (ed.), *Programmed Necrosis: Methods and Protocols*, Methods in Molecular Biology, vol. 1857, https://doi.org/10.1007/978-1-4939-8754-2_16, © Springer Science+Business Media, LLC, part of Springer Nature 2018

tography before detection in the mass analyzer. Energy is then applied to cleave the peptides specifically along the amide backbone to produce fragment ions. The resulting peptide mass fingerprints (PMFs) are subsequently subjected to in silico database searching to assign the peptide sequences and corresponding proteins.

To specifically identify and characterize components of a given complex, a sufficient purification process of said complex prior to assessment in the mass analyzer is crucial. Furthermore, insufficient sample purity will result in a higher background noise emanating from peptides originating from more abundant proteins present in the cell lysate. This would suppress signals from far less abundant but specific complex components. A single purification step is often not sufficient to achieve satisfactory purification.

Here, we describe an approach employing a specialized two-step immunoprecipitation, a so-called tandem affinity purification, for the purification of the tumor necrosis factor receptor 1 signaling complex (TNFR1-SC) and the subsequent steps necessary for mass spectrometry analysis. For this, the ligand TNF was fused to a modified version of the original tandem affinity purification (moTAP) tag [2, 3]. The moTAP tag consists of a 3× Flag epitope followed by a PreScission cleavage site and a 2× Strep epitope, enabling a combination of a first high-affinity purification followed by an elution step via a site-specific protease and a second affinity purification step. Using this approach, we were able to identify and characterize novel components of the TNFR1-SC [4, 5].

This technique is a sensitive and selective method to purify protein signaling complexes formed under physiological conditions and is not limited to the TNFR1-SC as the tag can be used with any ligand or intracellular protein.

2 Materials

Prepare all solutions and reagents strictly while wearing gloves to avoid protein contamination with keratin or other impurities from contact with skin (*see* **Note 1**).

2.1 Affinity Purification

1. moTAP-TNF. Sequence coding for moTAP-TNF was cloned into pQE-30 and expressed in BL21 *E. coli* by IPTG induction.
2. Lysis buffer: 30 mM Tris–HCl; pH 7.4, 120 mM NaCl, 2 mM EDTA, 2 mM KCl, 10% glycerol, 1% lauryl maltoside, 50 mM NaF, 5 mM Na_3VO_4. Store at 4 °C (*see* **Note 2**).
3. Protease inhibitors: cOmplete™, EDTA-free Protease Inhibitor Cocktail (Roche).
4. Phosphatase inhibitors: PhosSTOP™ (Sigma-Aldrich).
5. Anti-Flag resin: ANTI-FLAG® M2 Affinity Gel (Sigma-Aldrich).

6. Wash buffer: 30 mM Tris–HCl; pH 7.4, 120 mM NaCl, 2 mM EDTA, 2 mM KCl, 10% glycerol, 0.1% lauryl maltoside, 50 mM NaF, 5 mM Na_3VO_4. Store at 4 °C.
7. Microcentrifuge tubes, protein LoBind/DNA LoBind (Eppendorf).
8. 3× FLAG® Peptide (Sigma-Aldrich).
9. PreScission Protease (GE Healthcare).
10. Elution Buffer I (Flag peptide + PreScission protease): Wash buffer, 200 μg/mL 3× FLAG® Peptide, 50 U/mL PreScission protease.
11. Strep-Tactin Superflow Plus (Qiagen).
12. Elution Buffer II (urea): 100 mM ammonium bicarbonate, 8 M urea.

2.2 Protein Digestion

1. Digestion buffer: 100 mM ammonium bicarbonate.
2. Reduction solution: digestion buffer (100 mM ammonium bicarbonate), 200 mM dithiothreitol (DTT).
3. Alkylation solution: digestion buffer (100 mM ammonium bicarbonate), 200 mM chloroacetic acid (*see* **Note 3**).
4. Lysyl endopeptidase/Lys-C (Alpha Laboratories). Use 1:50–1:100 protease to protein ratio.
5. Trypsin–EDTA (Invitrogen). Use 1:50–1:100 protease to protein ratio.

2.3 C18 Cleanup

1. UltraMicro spin columns, Silica C18, 3–30 μg; 2–100 μL capacity (Part No: SUM SS18V), Nest Group Inc., Southborough, MA.
2. 80% acetonitrile, 0.1% trifluoroacetic acid (TFA).
3. 0.1% TFA.
4. 10% TFA.
5. 5% acetonitrile, 0.1% TFA.
6. 50% acetonitrile, 0.1% TFA.
7. 0.1% formic acid.

3 Methods

3.1 Cloning of moTAP Tagged Protein of Interest

The different parts of the moTAP-tag were ordered as complementary oligonucleotides and assembled sequentially into the pE30 plasmid using conventional restriction/ligation techniques. However, it is advisable to have the full tag commercially synthesized as dsDNA. This is offered by many companies and minimizes the potential for cloning mistakes. The amino acid sequence of the moTAP-TNF is shown below and a schematic depiction can be found in Fig. 1 (grey sequences

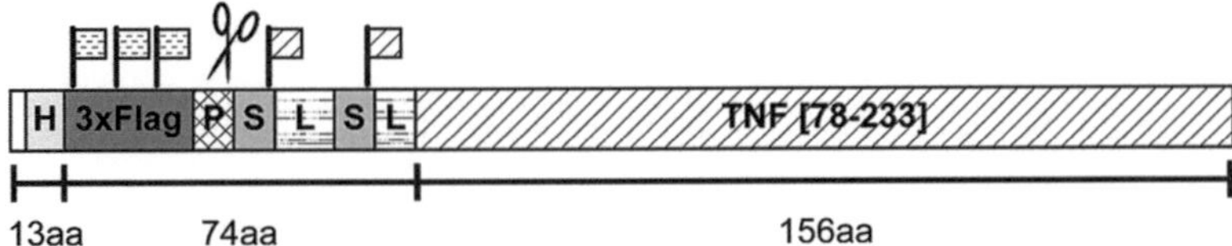

Fig. 1 Sequence of moTAP-TNF with annotation of different tag and cleavage site regions. H = His-Tag; P = PreScission cleavage site; S = Strep-Tag; L = Linker sequence

represent functional domains as depicted in Fig. 1; the sequence in italic represents the start of aa78–233 of TNF).

MRGS HHHHHH GIQ DYKDDDDK A DYKDDDDK AL DYKDDDDK A LEVLFQGP EL

WSHPQFEK GGGSGGGSGGGS WSHPQFEK GGGSTSRS *RSSSRTP ...*

3.2 Cell Lysis

The following instructions are per sample of cells grown to confluency in one 15 cm dish. Cell lysates must be strictly kept at 4 °C.

1. Stimulate cells with 500 ng/mL of moTAP-TNF.
2. Lyse cells using 1 mL lysis buffer containing protease inhibitors and phosphatase inhibitors and use cell scraper to transfer cell lysate into an Eppendorf tube (*see* **Note 4**).
3. Centrifuge cell lysate at 15,870 × *g* at 4 °C for 30 min to remove insoluble debris.
4. Transfer supernatant into a new Eppendorf tube.

3.3 First Purification Step (Flag-IP) and Elution from Anti-Flag Resin

1. Add 50 μL of prewashed, undiluted anti-Flag resin to the cell lysate and incubate overnight at 4 °C while keeping the tube in motion, ideally on a rotating wheel.
2. Centrifuge the anti-Flag resin at 2350 × *g* at 4 °C for 5 min and discard supernatant.
3. Wash the resin with 1 mL of wash buffer containing protease inhibitors and phosphatase inhibitors and repeat centrifugation step. Repeat the washing step two times and in the last step transfer the resin into a protein LoBind Eppendorf tube (*see* **Note 5**).
4. Elute the bound complex from the anti-Flag resin in 400 μL of Elution buffer I containing protease inhibitors and phosphatase inhibitors for 6 h at 4 °C.
5. Centrifuge the anti-Flag resin at 2350 × *g* at 4 °C for 2 min and transfer eluate into new protein LoBind Eppendorf tube.
6. Repeat elution step as above and incubate overnight at 4 °C while keeping the tube in motion.
7. Centrifuge the sample at 2350 × *g* at 4 °C for 2 min and transfer supernatant to merge with previously obtained eluate.

3.4 Second Purification Step (Strep-IP) and Elution from Strep-Tactin Resin

1. Add 60 μL of prewashed, undiluted Strep-Tactin resin to the eluate and incubate overnight at 4 °C while keeping the tube in motion.
2. Centrifuge the Strep-Tactin resin at 2350 × *g* at 4 °C for 5 min and discard supernatant.
3. Wash the resin with 1 mL of wash buffer containing protease inhibitors and phosphatase inhibitors and repeat centrifugation step. Repeat the washing step two times.
4. Elute bound complex in 120 μL of Elution buffer II at 37 °C for 30 min (*see* **Note 6**).
5. Centrifuge the sample at 2350 × *g* at 4 °C for 2 min and transfer eluate into new protein LoBind Eppendorf tube.

3.5 Protein Digestion

1. Add 2 μL of reduction solution to the eluate and incubate at 56 °C for 25 min. Afterward, let the sample cool down to room temperature.
2. Add 4 μL of alkylation solution to the sample and incubate at room temperature for 25 min, avoiding exposure to light sources.
3. Add 2 μL of reduction solution to neutralize the excess of chloroacetic acid.
4. Add lysyl endopeptidase in a 1:50–1:100 protease to protein ratio and incubate at 37 °C for 4 h. Afterward, let the sample cool down to room temperature (*see* **Note 7**).
5. Dilute the sample four times to a concentration of 2 M urea by adding denaturation buffer (~300 μL).
6. Add trypsin–EDTA in a 1:50–1:100 protease to protein ratio and incubate overnight at 37 °C. Afterward, let the sample cool down to room temperature.

3.6 C18 Cleanup

For C18 cleanup, use a micro spin column with a capacity of 3–30 μg of total protein and 2–100 μL of volume.

1. Activate the column by washing it with 200 μL of 80% acetonitrile–0.1% TFA while centrifuging the column at 100 × *g* at room temperature for 1 min for a total of two times.
2. Equilibrate column with 200 μL of 0.1% TFA for a total of three times.
3. Acidify samples to be loaded by adding 50 μL of 10% TFA, which will result in a final concentration of 1% TFA. Make sure that a pH of 2.5 is reached.
4. Centrifuge the acidified samples at 15,870 × *g* at room temperature for 5 min.
5. Load only the supernatant of the acidified sample on the activated column and centrifuge the column at 100 × *g* at room

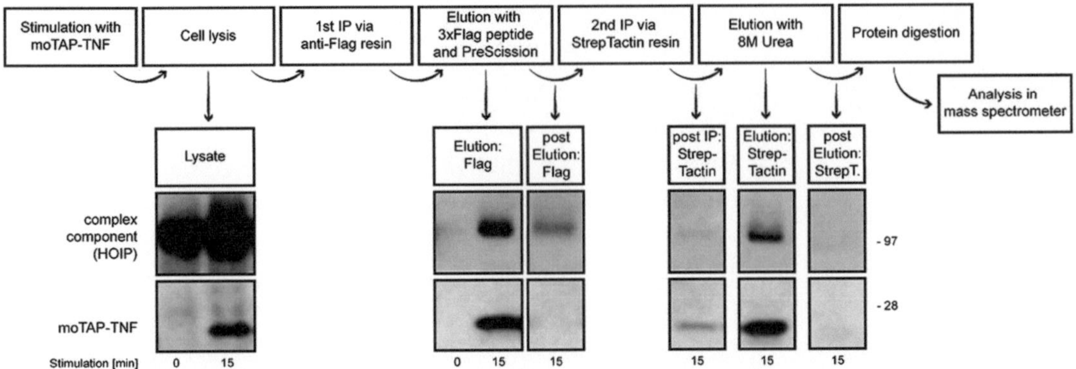

Fig. 2 Flowchart of steps described in the protocol. Aliquots of lysate and samples after each purification step were analyzed by western blot with subsequent staining of moTAP-TNF and the representative complex component HOIP. IP=Immunoprecipitation; post = Supernatant remaining after respective purification step; StrepT = Strep-Tactin

temperature for 1 min. Collect the flow-through in a clean tube. This step might have to be repeated for the whole supernatant to run through the column.

6. Reload the sample flow-through and repeat the centrifugation step as above to maximize amount of protein binding to the column.
7. Wash the column with 200 μL of 5% acetonitrile–0.1% TFA to wash away and dilute urea for a total of five times.
8. Place a new tube to collect the eluate and elute the samples with 50 μL 50% acetonitrile–0.1% TFA while centrifuging the column at 100 × *g* at room temperature for 1 min for a total of two times.
9. Concentrate peptides in the eluate to dryness at 23 °C, ideally using speed vacuum centrifugation at 2000 rpm at room temperature (*see* **Note 8**).
10. The lyophilized peptides should be stored at −80 °C until ready to run on the LC-MS.
11. Once ready to run on the mass spectrometer, resuspend the peptides in 10 μL 0.1% formic acid and proceed with analysis (*see* **Note 9**). Figure 2 shows the work flow for the isolation of the receptor complex by moTAP together with the analysis of samples from each step.

3.7 Peptide Detection by LC-MS Analysis

Reversed phase chromatography using C18 particles is highly advised for peptide detection.

1. Peptides are concentrated on a precolumn C18 trap.
2. Peptides are then resolved using a C18 analytical column over 60 min with a linear gradient of 96:4 to 50:50 buffer A:B (buf-

fer A: 0.1% formic acid in water; buffer B: 0.1% formic acid in acetonitrile) between 200–300 nL/min.

3. Peptides are then ionized by electrospray ionization (ESI) before being infused into the mass spectrometer. A specific type of data acquisition, known as data-dependent acquisition (DDA) is then triggered.
4. Firstly, the peptides are measured in the mass analyzer with high accuracy and resolution power. MS/MS events are then acquired by selecting the most intense ions to be fragmented by (higher-)collision-induced dissociation (HCD or CID). The precise number of peptide ions sequenced at a given time is determined by the instruments duty cycle.
5. The fragment ions of a particular peptide are then reanalyzed in the mass analyzer to determine its PMF.
6. Once fully acquired, PMF data can be searched against the appropriate databases in silico using search engines such as Mascot, Sequest, and MaxQuant.

4 Notes

1. It is of great importance to always wear gloves while handling samples and preparing buffers and solutions. For weighing chemicals, special dedicated instruments should be used that have not been touched with bare hands and should be cleaned before use. Filtered pipette tips should be used. This is to prevent contamination of the mass spec samples with keratins or other proteins present in the skin that would cause a contamination of the sample.
2. The detergent Triton must not be used for sample preparation or in any solution that will be subjected to LC-MS analysis. Even small traces of Triton will render samples incompatible for mass spectrometry.
3. If the ubiquitination status of proteins is going to be analyzed, we recommend the use of chloroacetic acid over iodoacetamide in the preparation of alkylation solution. This is because alkylation by iodoacetamide can mimic the di-gly remnant which is indicative for protein ubiquitination [6].
4. For optimization purposes and troubleshooting it is highly recommended to take aliquots of samples after each purification step for subsequent analysis by western blot. This should be done at the initial step of lysis, for the supernatants that are being discarded after the respective immunoprecipitations and for the eluates. Furthermore, the affinity resins should be analyzed after the respective elution steps for remainders of the complex of interest

that was not successfully eluted. This analysis allows to identify inefficient purification steps that must be optimized.

5. It is recommended to use protein LoBind tubes from the point of elution from the anti-Flag resin onward. This will minimize the loss of protein which results from unspecific binding to the surface of conventional reaction tubes.
6. The elution from the Strep-Tactin resin using urea is an efficient yet somewhat crude approach. If suspicions of unspecific binding of proteins to the Strep-Tactin resin should arise, it is recommended to use a more specific approach for elution. An alternative is therefore eluting with denaturation buffer and 5 mM biotin at 4 °C for 24 h with subsequent protein precipitation (using "2-D Clean-Up Kit" (GE Healthcare) according to manual).
7. The use of LysC in conjunction with trypsin enhances digestion by reducing the tryptic missed cleavage rate. LysC is able to function in harsh conditions such as 8 M urea, which means that it is able to cleave fully denatured proteins efficiently. Trypsin is then introduced to cleave these long peptides after internal arginine residues.
8. Before the final step of the C18 cleanup, the peptides contained in the samples should ideally be concentrated to dryness leaving a small droplet at 23–30 °C using a vacuum centrifuge (e.g., ScanVac). When using a vacuum centrifuge, take precautions not to contaminate samples with dirt and spillage remainders present in the centrifuge, as reaction tubes must stay opened during centrifugation. Clean the centrifuge prior to use if necessary.
9. Thorough resolubilization of peptides is carried out by vortexing followed by sonicating for 10 min. Peptides should be left for at least 30 min to allow for maximum resolubilization.

Acknowledgments

We thank P. Draber and L. Spilgies for providing crucial technical advice and L. Spilgies for providing western blot stainings. This work was supported by a Wellcome Trust Senior Investigator Award (096831/Z/11/Z), an ERC Advanced Grant (294880), and a Cancer Research UK programme grant (A17341) awarded to H.W.; and a BBSRC CASE studentship (BB/J013129/1) awarded to M.R. This work was also supported by the CRUK–UCL Centre grant (515818) and the Cancer Immunotherapy Accelerator (CITA) Award (525877) from CRUK.

References

1. Aebersold R, Mann M (2016) Mass-spectrometric exploration of proteome structure and function. Nature 537(7620):347–355. https://doi.org/10.1038/nature19949
2. Haas TL, Emmerich CH, Br G, Schmukle AC, Cordier SM, Rieser E, Feltham R, Vince J, Warnken U, Wenger T, Koschny R, Komander D, Silke J, Walczak H (2009) Recruitment of the linear ubiquitin chain assembly complex stabilizes the TNF-R1 signaling complex and¬†is required for TNF-mediated gene induction. Mol Cell 36(5):831–844. https://doi.org/10.1016/j.molcel.2009.10.013
3. Rigaut G, Shevchenko A, Rutz B, Wilm M, Mann M, Seraphin B (1999) A generic protein purification method for protein complex characterization and proteome exploration. Nat Biotechnol 17(10):1030–1032. https://doi.org/10.1038/13732
4. Draber P, Kupka S, Reichert M, Draberova H, Lafont E, de Miguel D, Spilgies L, Surinova S, Taraborrelli L, Hartwig T, Rieser E, Martino L, Rittinger K, Walczak H (2015) LUBAC-recruited CYLD and A20 regulate gene activation and cell death by exerting opposing effects on linear ubiquitin in signaling complexes. Cell Rep 13(10):2258–2272. https://doi.org/10.1016/j.celrep.2015.11.009
5. Kupka S, De Miguel D, Draber P, Martino L, Surinova S, Rittinger K, Walczak H (2016) SPATA2-mediated binding of CYLD to HOIP enables CYLD recruitment to signaling complexes. Cell Rep 16(9):2271–2280. https://doi.org/10.1016/j.celrep.2016.07.086
6. Nielsen ML, Vermeulen M, Bonaldi T, Cox J, Moroder L, Mann M (2008) Iodoacetamide-induced artifact mimics ubiquitination in mass spectrometry. Nat Methods 5(6):459–460. https://doi.org/10.1038/nmeth0608-459

Chapter 17

Monitoring RIPK1 Phosphorylation in the TNFR1 Signaling Complex

Dario Priem, Yves Dondelinger, and Mathieu J. M. Bertrand

Abstract

Receptor-interacting protein kinase 1 (RIPK1) is a component of the TNFR1 signaling complex (also known as complex I or TNFR-SC), where its ubiquitylation by cIAP1/2 and LUBAC serves to initiate prosurvival and proinflammatory responses through activation of the MAPK and NF-κB pathways. IKKα/β-mediated phosphorylation of RIPK1 in complex I was shown to maintain RIPK1 in a prosurvival modus. Consequently, conditions affecting proper IKKα/β activation perturb IKKα/β-phosphorylation of RIPK1 and switch the TNF response toward RIPK1 kinase-dependent cell death. Methods to study the posttranslational modifications of RIPK1 in complex I are therefore of great value. Here, we describe a detailed protocol to isolate complex I-associated RIPK1 from cells and provide different tools to study the phosphorylation status of RIPK1 in TNFR1 complex I.

Key words TNF, TNFR1, Complex I, RIPK1, Phosphorylation, Ubiquitylation, Cell death, Immunoprecipitation, Immunoblotting, Necroptosis, Apoptosis

1 Introduction

Tumor necrosis factor (TNF) is a master proinflammatory cytokine whose pathogenic role in inflammatory disorders can, in certain conditions, be attributed to receptor-interacting protein kinase 1 (RIPK1)-dependent cell death, in the form of apoptosis or necroptosis [1–4]. Cell survival is however the default response of most cells to TNFR1 activation, indicating that cell demise is normally actively repressed and that certain cell death checkpoints need to be disrupted for cell death to proceed [5].

TNFR1 engagement by TNF triggers the recruitment of RIPK1 to a plasma membrane-bound signaling complex, known as complex I or TNFR1-SC, where RIPK1 functions as a scaffold to activate the MAPK and NF-κB signaling pathways [6]. The conjugation of ubiquitin chains by cIAP1/2 and LUBAC (composed of HOIP, HOIL-1 and SHARPIN) to RIPK1, and probably other complex I components, generates binding sites for the adaptor proteins

Adrian T. Ting (ed.), *Programmed Necrosis: Methods and Protocols*, Methods in Molecular Biology, vol. 1857,
https://doi.org/10.1007/978-1-4939-8754-2_17,

TAB2/3 and NEMO and allows further recruitment and activation of the kinases TAK1 and IKKα/β [7–13]. Activation of IKKβ by TAK1 then leads to IKKβ-dependent phosphorylation and subsequent proteasomal degradation of IκBα, and consequently to the translocation of the NF-κB heterodimer p50/p65 to the nucleus where it drives transcription of multiple responsive genes.

Inhibition of NF-κB-dependent transcription is known to switch the TNF response and induce limited apoptosis through assembly of a cytosolic caspase-8 activating platform depending on TRADD and FADD and known as complex IIa [14, 15]. The NF-κB-dependent expression of prosurvival molecules counteracting complex IIa assembly/activation therefore represents a late cell death checkpoint in the TNFR1 pathway. An additional transcription-independent cell death checkpoint taking place at the level of complex I and controlling the contribution of RIPK1 to the cell demise was more recently reported [5, 16]. The molecular nature of this early cell death checkpoint is not yet fully understood but shown to be controlled by IKKα/β phosphorylation of RIPK1 [2]. Consequently, any conditions affecting proper IKKα/β activation in complex I (such as cIAP1/2, NEMO, or SHARPIN depletion and TAK1 or IKKα/β inhibition) results in defective phosphorylation of RIPK1 and induction of RIPK1 kinase-dependent cell death [1–3, 15, 17, 18]. Under these conditions, the caspase-8 activation platform does not rely on TRADD but instead on RIPK1 and its enzymatic activity [15]. This alternative complex II has therefore been defined as complex IIb [19]. Phosphorylation of RIPK1 by IKKα/β in complex I was also shown to regulate TNFR1-mediated RIPK1 kinase-dependent necroptosis [2, 18, 20]. Monitoring RIPK1 phosphorylation in TNFR1 complex I can therefore serve as an important readout to evaluate the status of the early RIPK1 cell death checkpoint, and for the identification of new regulators of RIPK1 kinase-dependent death.

2 Materials

1. Recombinant human FLAG-tagged human TNF (FLAG-TNF).
2. Cell culture medium: depending on the mouse cell type.
3. Phosphate-buffered saline (PBS): 137 mM NaCl, 2.7 mM KCl, 8 mM Na_2HPO_4, 1.5 mM KH_2PO_4. Store at 4 °C.
4. Cell scrapers.
5. ANTI-FLAG® M2 Affinity Gel (Sigma).
6. NP-40 buffer: 150 mM NaCl, 1% NP-40, 10% glycerol, 10 mM Tris–HCl pH 8. Store at 4 °C.

7. Lysis buffer: NP-40 buffer containing cOmplete™ Protease Inhibitor Cocktail (Roche) and PhosSTOP™ (Roche) phosphatase inhibitor. Store at 4 °C (*see* **Note 1**).
8. Laemmli buffer (5×): 10% (w/v) SDS, 50% glycerol, 0.5% bromophenol blue, 250 mM Tris–HCl pH 6.8. Store aliquots at −20 °C. Add dithiothreitol (DTT) at 250 mM right before use (*see* **Note 2**).
9. DUB buffer: 50 mM NaCl, 50 mM Tris–HCl pH 8. Add DTT at 5 mM right before use (*see* **Note 2**).
10. DUB/λPPase buffer: 1 mM $MnCl_2$, 50 mM NaCl, 50 mM Tris–HCl pH 8. Add DTT at 5 mM right before use (*see* **Note 2**).
11. Recombinant rat USP2 (Enzo).
12. Lambda Protein Phosphatase (λPPase) (Bioké).

3 Methods

3.1 TNFR1 Complex I Pull-Down

1. For each condition, seed the mouse cells of interest in a cell culture dish (⌀ 10 cm) to obtain a fully confluent plate the next day. In this chapter, 2.5 × 10^6 mouse embryonic fibroblasts (MEFs) were seeded in 10 mL of cell culture medium (*see* **Note 3**).
2. The next day, the cells are either left unstimulated or stimulated with 1 μg/mL FLAG-tagged human TNF in 5 mL of medium for increasing periods of time (5–60 min). It is recommended to optimize the concentration of FLAG-TNF for different cell lines (Fig. 1) (*see* **Note 4**).
3. To stop the stimulation, aspirate the medium and wash the cells twice with 10 mL ice-cold PBS. Aspirate all remaining PBS and keep the cells on ice from that moment on.
4. Add 1 mL of lysis buffer to each cell culture dish and incubate for 15 min on a rocking table at 4 °C.
5. Use a cell scraper to harvest each cell lysate and transfer them to 1.5 mL Eppendorf tubes. Keep the lysates on ice.
6. Clear the lysate by centrifugation for 10 min at max speed (>10,000 × *g*) in a microcentrifuge at 4 °C.
7. In the meantime, prepare the ANTI-FLAG M2 Affinity Gel. Use 30 μL of the 50% beads solution per immunoprecipitation (*see* **Note 5**). Wash the beads twice with 1 mL of cold NP-40 buffer. Between washes, centrifuge the beads for 1 min at 500 × *g* at 4 °C and remove the supernatant carefully (*see* **Note 6**). Keep the beads on ice.
8. After centrifugation, collect 50 μL of the cleared cell lysates in separate Eppendorf tubes and add 12.5 μL 5× Laemmli buffer

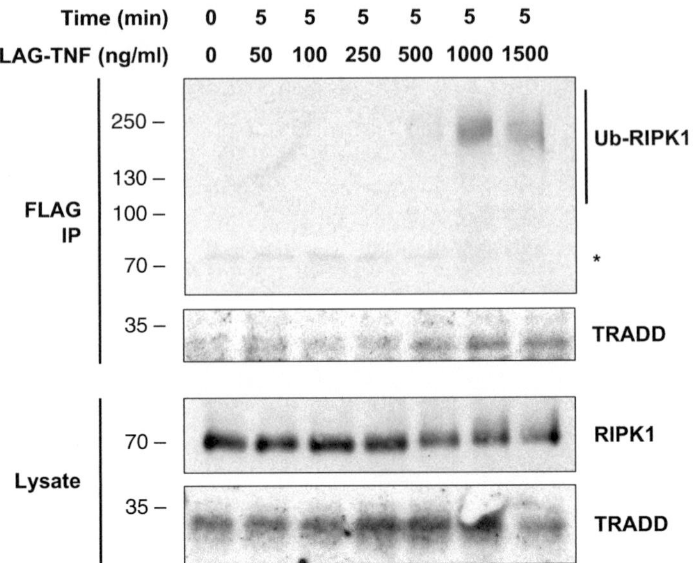

Fig. 1 Mouse embryonic fibroblasts (MEFs) were stimulated with different concentrations of recombinant FLAG-tagged human TNF for 5 min or left unstimulated. TNFR1 complex I was immunoprecipitated using the method described in this chapter. Recruitment of RIPK1 and TRADD to TNFR1 complex I was determined by immunoblotting. Ubiquitylated RIPK1 appears as a smear. * represents a nonspecific band

to them. They will form the **pre-IP** lysate samples. Boil the samples for 5 min at 95 °C and use them for Western blot analysis or store them at −20 °C for later analysis.

9. Transfer the rest of the supernatants to the Eppendorf tubes containing the prewashed ANTI-FLAG M2 gel.
10. Incubate the supernatants with the ANTI-FLAG M2 gel (3 h to overnight) at 4 °C on a spinning wheel.
11. Centrifuge the sample for 1 min at 500 × *g* at 4 °C.
12. Optional: collect 50 μL of the supernatants in separate Eppendorf tubes and add 12.5 μL 5× Laemmli buffer. They will constitute the **post-IP** lysate samples (*see* **Note 7**). Boil the sample for 5 min at 95 °C and use for Western blot analysis or store at −20 °C for later analysis.
13. Remove the rest of the supernatant (*see* **Note 6**) and wash the beads with 1 mL of cold NP-40 buffer. Centrifuge for 1 min at 500 × *g* at 4 °C and remove the supernatant carefully. Repeat the washing step 3 times. After the last wash, centrifuge the samples for 1 min at 500 × *g* at 4 °C and remove the remaining buffer with a pipette and a protein loading tip.

14. If a USP2 or λ phosphatase (λPPase) treatment is required, proceed to the next step. If not, resuspend the beads in 60 μL 1× Laemmli buffer. This constitutes the **IP** sample. Boil the sample for 5 min at 95 °C and use for Western blot analysis or store at −20 °C for later analysis.

3.2 USP2 and λPPase Treatment

1. If only an USP2 treatment is required: resuspend the immunoprecipitate in 50 μL DUB buffer and add 500 ng of recombinant USP2 (*see* **Note 8**). It is recommended to optimize the concentration of USP2 depending on the amount of protein that you immunoprecipitate (Fig. 2).
2. If an additional λPPase treatment is required: resuspend the beads in 50 μL DUB/λPPase buffer, add 500 ng of recombinant USP2 and 800 U of recombinant λPPase (Fig. 3) (*see* **Note 9**).
3. Incubate the sample for 30 min at 37 °C.
4. To terminate the reaction, add 12.5 μL 5× Laemmli buffer. Boil the sample for 5 min at 95 °C and use for Western blot analysis or store at −20 °C for later analysis.

4 Notes

1. Prepare this buffer freshly before each experiment to ensure optimal efficiency of the inhibitor cocktail.
2. First dissolve the SDS before adding the glycerol, otherwise it will not dissolve. Heating the solution can improve the solubility. DTT is very unstable when dissolved and multiple freeze–thaw cycles should be avoided.
3. In principle, every type of mouse cell can be used as long as the cells express enough TNFR1 to be pulled down. Adjust the number of cells depending on the cell type.
4. In order to avoid immunoprecipitating both TNFR1 and TNFR2, we use human TNF on mouse cells. Indeed, in contrast to mouse TNF that binds both mouse TNFR1 and TNFR2, human TNF only interacts with mouse TNFR1. To reduce the amount of recombinant FLAG-TNF needed per experiment, it is recommended to reduce the volume of culture medium prior to stimulation. In a 10 cm cell culture dish, you can reduce the volume to 5 mL, allowing the cells still to be fully covered by culture medium.
5. ANTI-FLAG M2 Affinity gel is stored as a 50% solution. It is important to mix the solution thoroughly before use.
6. Avoid losing gel during the removal of the supernatant. The majority of the supernatant can be removed by aspiration, but avoid getting too close to the beads. Remove the rest of the supernatant carefully using a pipet and a protein loading tip.

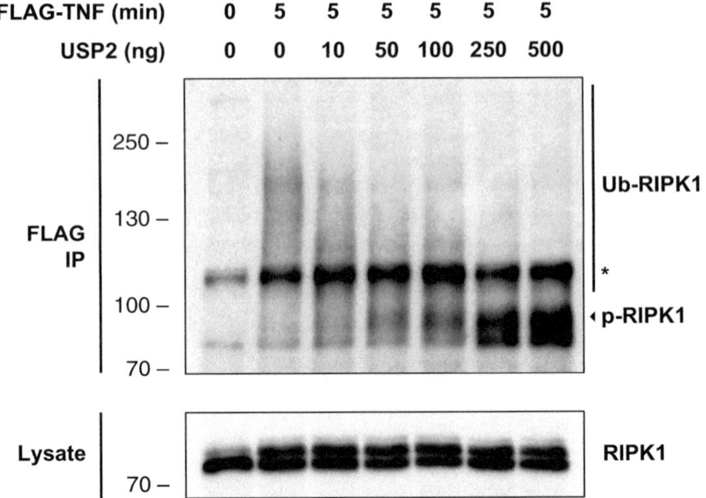

Fig. 2 MEFs were stimulated with 1 μg/mL FLAG-tagged human TNF or left unstimulated. After immunoprecipitation of complex I, the samples were treated with different concentrations of recombinant USP2. The removal of ubiquitin chains from RIPK1 was monitored by immunoblotting for RIPK1. Removal of the ubiquitin chains by USP2 treatment reveals a RIPK1 signal at a molecular weight that is higher than the one of unmodified RIPK1. * represents a nonspecific band

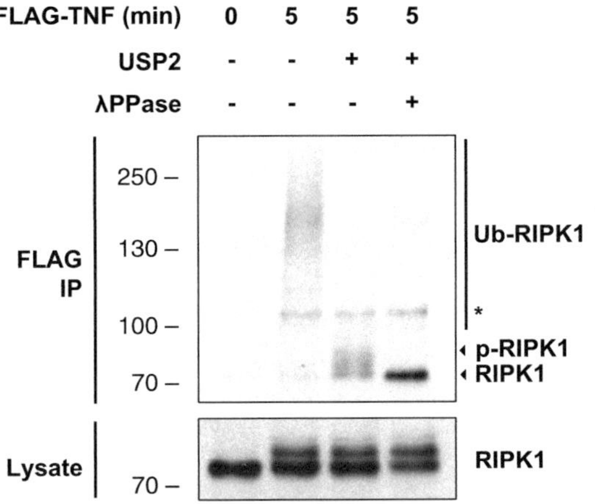

Fig. 3 MEFs were stimulated with 1 μg/mL FLAG-tagged human TNF or left unstimulated. After complex I immunoprecipitation, the samples were treated with 500 ng USP2 in presence or absence of λPPase, and recruitment of RIPK1 to complex I was monitored by immunoblotting for RIPK1. The λPPase treatment demonstrates that the RIPK1 signal revealed by USP2 deubiquitylation corresponds to a mobility shift caused by phosphorylation. * represents a nonspecific band

7. The post-IP sample can be used to check the efficiency of the pull-down reaction. Pre- and post-IP lysates can by immunoblotted with anti-FLAG antibodies to evaluate the efficiency of the pull-down. The post-IP supernatant can also be used for a second immunoprecipitation reaction, e.g., to immunoprecipitate the pool of RIPK1 that is not present in TNFR1 complex I.
8. RIPK1 and other components of the TNFR1 signaling complex are ubiquitylated [9–11, 13], and therefore appear as smears when analyzed by immunoblot. In order to detect a mobility shift of RIPK1 resulting from its phosphorylation, the ubiquitin chains conjugated to RIPK1 have to be removed. This is obtained by treating the immunoprecipitated complex I with USP2, a deubiquitylase (DUB) known to cleave all types of ubiquitin chains [21]. USP2 treatment results in the collapse of the ubiquitylated smear and allows detection of the mobility shift resulting from IKK-mediated phosphorylation of RIPK1 (Fig. 2).
9. IKK-phosphorylated RIPK1 is distinguishable from unmodified RIPK1 by a mobility shift detected on immunoblot (Fig. 3). The mobility shift caused by phosphorylation of RIPK1 is best detected on an 8% bis-acrylamide gel. To confirm that the mobility shift is indeed caused by the phosphorylation of RIPK1, a λPPase treatment can be performed to fully dephosphorylate RIPK1. This mobility shift can also be inhibited by blocking the activation of IKKs or of a kinase acting upstream in the pathway, such as TAK1 (Fig. 4) [2].

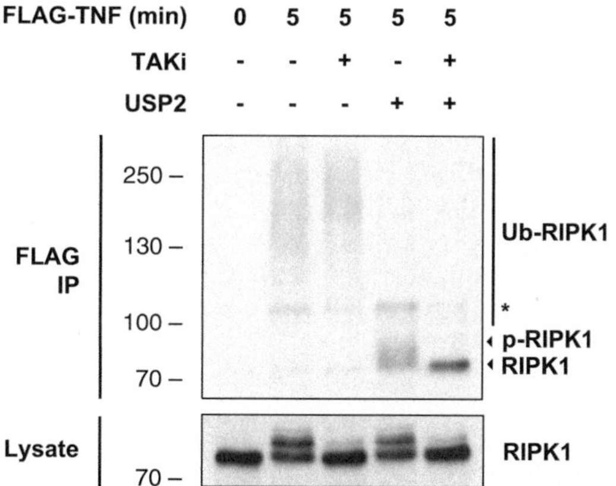

Fig. 4 MEFs were pretreated with TAK1 inhibitor (TAKi) or DMSO for 30 min before stimulating the cells with 1 μg/mL FLAG-tagged human TNF. Complex I was immunoprecipitated and treated or not with 500 ng of USP2. Recruitment of RIPK1 to complex I was monitored by immunoblot. Inhibition of TAK1, a kinase acting upstream of IKKs, prevents the mobility shift caused by IKK-mediated phosphorylation of RIPK1 that is detected upon RIPK1 deubiquitylation by USP2. * represents a nonspecific band

References

1. Berger SB, Kasparcova V, Hoffman S, Swift B, Dare L, Schaeffer M, Capriotti C, Cook M, Finger J, Hughes-Earle A, Harris PA, Kaiser WJ, Mocarski ES, Bertin J, Gough PJ (2014) Cutting edge: RIP1 kinase activity is dispensable for normal development but is a key regulator of inflammation in SHARPIN-deficient mice. J Immunol 192(12):5476–5480. https://doi.org/10.4049/jimmunol.1400499
2. Dondelinger Y, Jouan-Lanhouet S, Divert T, Theatre E, Bertin J, Gough PJ, Giansanti P, Heck AJ, Dejardin E, Vandenabeele P, Bertrand MJ (2015) NF-kappaB-independent role of IKKalpha/IKKbeta in preventing RIPK1 kinase-dependent apoptotic and necroptotic cell death during TNF signaling. Mol Cell 60(1):63–76. https://doi.org/10.1016/j.molcel.2015.07.032
3. Vlantis K, Wullaert A, Polykratis A, Kondylis V, Dannappel M, Schwarzer R, Welz P, Corona T, Walczak H, Weih F, Klein U, Kelliher M, Pasparakis M (2016) NEMO prevents RIP kinase 1-mediated epithelial cell death and chronic intestinal inflammation by NF-kappaB-dependent and -independent functions. Immunity 44(3):553–567. https://doi.org/10.1016/j.immuni.2016.02.020
4. Newton K, Wickliffe KE, Maltzman A, Dugger DL, Strasser A, Pham VC, Lill JR, Roose-Girma M, Warming S, Solon M, Ngu H, Webster JD, Dixit VM (2016) RIPK1 inhibits ZBP1-driven necroptosis during development. Nature 540(7631):129–133. https://doi.org/10.1038/nature20559
5. Ting AT, Bertrand MJ (2016) More to life than NF-kappaB in TNFR1 signaling. Trends Immunol 37(8):535–545. https://doi.org/10.1016/j.it.2016.06.002
6. Brenner D, Blaser H, Mak TW (2015) Regulation of tumour necrosis factor signalling: live or let die. Nat Rev Immunol 15(6):362–374. https://doi.org/10.1038/nri3834
7. Kanayama A, Seth RB, Sun L, Ea CK, Hong M, Shaito A, Chiu YH, Deng L, Chen ZJ (2004) TAB2 and TAB3 activate the NF-kappaB pathway through binding to polyubiquitin chains. Mol Cell 15(4):535–548. https://doi.org/10.1016/j.molcel.2004.08.008
8. Ea CK, Deng L, Xia ZP, Pineda G, Chen ZJ (2006) Activation of IKK by TNFalpha requires site-specific ubiquitination of RIP1 and polyubiquitin binding by NEMO. Mol Cell 22(2):245–257
9. Bertrand MJ, Milutinovic S, Dickson KM, Ho WC, Boudreault A, Durkin J, Gillard JW, Jaquith JB, Morris SJ, Barker PA (2008) cIAP1 and cIAP2 facilitate Cancer cell survival by functioning as E3 ligases that promote RIP1 ubiquitination. Mol Cell 30(6):689–700
10. Varfolomeev E, Goncharov T, Fedorova AV, Dynek JN, Zobel K, Deshayes K, Fairbrother WJ, Vucic D (2008) c-IAP1 and c-IAP2 are critical mediators of tumor necrosis factor alpha (TNFalpha)-induced NF-kappaB activation. J Biol Chem 283(36):24295–24299. https://doi.org/10.1074/jbc.C800128200
11. Haas TL, Emmerich CH, Gerlach B, Schmukle AC, Cordier SM, Rieser E, Feltham R, Vince J, Warnken U, Wenger T, Koschny R, Komander D, Silke J, Walczak H (2009) Recruitment of the linear ubiquitin chain assembly complex stabilizes the TNF-R1 signaling complex and is required for TNF-mediated gene induction. Mol Cell 36(5):831–844. https://doi.org/10.1016/j.molcel.2009.10.013
12. Gerlach B, Cordier SM, Schmukle AC, Emmerich CH, Rieser E, Haas TL, Webb AI, Rickard JA, Anderton H, Wong WW, Nachbur U, Gangoda L, Warnken U, Purcell AW, Silke J, Walczak H (2011) Linear ubiquitination prevents inflammation and regulates immune signalling. Nature 471(7340):591–596. https://doi.org/10.1038/nature09816
13. Mahoney DJ, Cheung HH, Mrad RL, Plenchette S, Simard C, Enwere E, Arora V, Mak TW, Lacasse EC, Waring J, Korneluk RG (2008) Both cIAP1 and cIAP2 regulate TNFalpha-mediated NF-kappaB activation. Proc Natl Acad Sci U S A 105(33):11778–11783. https://doi.org/10.1073/pnas.0711122105
14. Van Antwerp DJ, Martin SJ, Kafri T, Green DR, Verma IM (1996) Suppression of TNF-alpha-induced apoptosis by NF-kappaB. Science 274(5288):787–789
15. Wang L, Du F, Wang X (2008) TNF-alpha induces two distinct caspase-8 activation pathways. Cell 133(4):693–703
16. O'Donnell MA, Legarda-Addison D, Skountzos P, Yeh WC, Ting AT (2007) Ubiquitination of RIP1 regulates an NF-kappaB-independent cell-death switch in TNF signaling. Curr Biol 17(5):418–424
17. Legarda-Addison D, Hase H, O'Donnell MA, Ting AT (2009) NEMO/IKKgamma regulates an early NF-kappaB-independent cell-death checkpoint during TNF signaling. Cell Death Differ 16(9):1279–1288. https://doi.org/10.1038/cdd.2009.41

18. Dondelinger Y, Aguileta MA, Goossens V, Dubuisson C, Grootjans S, Dejardin E, Vandenabeele P, Bertrand MJ (2013) RIPK3 contributes to TNFR1-mediated RIPK1 kinase-dependent apoptosis in conditions of cIAP1/2 depletion or TAK1 kinase inhibition. Cell Death Differ 20(10):1381–1392. https://doi.org/10.1038/cdd.2013.94
19. Wilson NS, Dixit V, Ashkenazi A (2009) Death receptor signal transducers: nodes of coordination in immune signaling networks. Nat Immunol 10(4):348–355. https://doi.org/10.1038/ni.1714
20. O'Donnell MA, Hase H, Legarda D, Ting AT (2012) NEMO inhibits programmed necrosis in an NFkappaB-independent manner by restraining RIP1. PLoS One 7(7):e41238. https://doi.org/10.1371/journal.pone.0041238
21. Hospenthal MK, Mevissen TET, Komander D (2015) Deubiquitinase-based analysis of ubiquitin chain architecture using ubiquitin chain restriction (UbiCRest). Nat Protoc 10(2):349–361. https://doi.org/10.1038/nprot.2015.018

Chapter 18

Analysis of CYLD Proteolysis by CASPASE 8 in Bone Marrow-Derived Macrophages

Diana Legarda and Adrian T. Ting

Abstract

Previous studies have demonstrated that CASPASE 8 can generate a prosurvival signal by inhibiting necroptosis via the cleavage of the deubiquitinating enzyme CYLD. Cleavage of CYLD at D215 results in the generation of a 25 kD N-terminal fragment and degradation of the C-terminal fragment containing the catalytic domain. Since CYLD is required for TNF-induced necroptosis, its proteolysis is necessary and sufficient to suppress necroptosis and generate a survival signal. Here we describe how to visualize CYLD proteolysis by western blot analysis, as a measure of CASPASE 8 activity and inhibition of necroptosis.

Key words Bone marrow-derived macrophages, Necroptosis, CYLD, Proteolysis, Cleavage

1 Introduction

Cylindromatosis, or CYLD, is a deubiquitinating enzyme that is required for tumor necrosis factor (TNF)-induced necroptosis [1–3]. It removes lysine-63-linked polyubiquitin chains from RIPK1, a required mediator of necroptosis. We have previously demonstrated that upon TNF receptor-1 (TNFR1) ligation, CYLD is cleaved at aspartate 215 (D215) by the prosurvival CASPASE 8. This cleavage results in the disappearance of the full-length 107 kDa CYLD (p107), the rapid degradation of the C-terminal fragment, and the appearance of an N-terminal, 25 kDa fragment (p25) that can be visualized by western blot analysis [2]. The proteolysis of CYLD by CASPASE 8 preserves RIPK1 ubiquitination, resulting in a survival response. Subsequently, we found that in macrophages, the engagement of Toll-like receptor 4 (TLR4) by lipopolysaccharide (LPS) also resulted in the cleavage of CYLD at D215 by CASPASE 8 [4]. TLR4 is able to directly activate CASPASE 8 to remove CYLD and disable the TNF necroptosis pathway [4]. Therefore, this CASPASE 8–CYLD interaction serves as a critical switch between survival and necroptosis.

Adrian T. Ting (ed.), *Programmed Necrosis: Methods and Protocols*, Methods in Molecular Biology, vol. 1857,
https://doi.org/10.1007/978-1-4939-8754-2_18,

Here we describe a method to detect the proteolysis of CYLD in mouse bone marrow-derived macrophages (BMDMs) upon stimulation with LPS. Lysates from BMDMs treated with LPS were subject to sodium dodecyl sulfate–polyacrylamide gel electrophoresis (SDS-PAGE), transferred to nitrocellulose membrane, and subject to western blot analysis using a commercially available CYLD monoclonal antibody that recognizes the N-terminus of CYLD. This protocol can be easily adapted to detect CYLD cleavage in response to other stimulation, e.g., TNF, in other cell types.

2 Materials

2.1 Cell Culture

1. Mouse BMDMs (*see* Chapter 6).
2. L929 conditioned media (L929-CM; *see* Chapter 6).
3. RPMI complete: RPMI 1640, 10% fetal calf serum (FCS), 100 IU penicillin 100 μg/mL streptomycin solution, 15 μg/mL gentamicin, 1% nonessential amino acids, and 0.1% beta-mercaptoethanol.
4. Humidified incubator at 37 °C with 5% CO_2.
5. Hemocytometer.
6. 0.25% trypsin containing 2.21 mM EDTA.
7. 1×PBS.
8. 24-well tissue culture treated plates.
9. 15 and 50 mL conical tubes.
10. Lipopolysaccharide (LPS) from *E. coli* 0111:B4.

2.2 SDS–Polyacrylamide Gel Electrophoresis (SDS-PAGE)

1. Cell lysis buffer: 20 mM Tris base, 40 mM NaCl, 5 mM EDTA, 50 mM NaF, 30 mM sodium pyrophosphate ($Na_4P_2O_7$), and 1% Triton X-100, and phosphatase inhibitors (such as Protease Inhibitor Cocktail Set V, EMD Millipore) (*see* **Note 1**). Adjust pH to 7.4.
2. Pierce BCA protein assay kit (ThermoFisher Scientific).
3. 5× protein loading buffer: 0.25% bromophenol blue, 0.5 M dithiothreitol (DTT), 50% glycerol, 10% sodium dodecyl sulfate (SDS), and 3.6 M beta-mercaptoethanol.
4. Heat block set to 95 °C.
5. Electrophoresis system, such as the Mini-Protean® system (Bio-Rad).
6. 1 M Tris–HCl, pH 8.8 for resolving gel.
7. 1 M Tris–HCl, pH 6.8 for stacking gel.

8. 40% acrylamide–bis solution in 37.5:1 ratio.
9. 10% sodium dodecyl sulfate (SDS).
10. N,N,N,N′-tetramethyl-ethylenediamine (TEMED).
11. 10% ammonium persulfate (APS) solution in distilled water, prepared fresh daily.
12. Water-saturated 1-butanol.
13. 10% resolving gel: 375 mM of Tris–HCl pH 8.8, 10% of acrylamide acrylamide:bis solution, 0.1% SDS, 0.1% APS, and 0.1% TEMED.
14. 5% stacking gel: 125 mM of Tris–HCl pH 6.8, 5% of acrylamide mix, 0.1% SDS, 0.1% APS, and 0.1% TEMED.
15. 10× Tris–glycine–SDS running buffer.
16. Prestained protein ladder, such as PageRuler™ prestained protein ladder 10 to 180 kDa (ThermoFisher Scientific).

2.3 Immunoblotting

1. Wet transfer system, such as the Mini Trans-Blot® electrophoretic transfer cell (Bio-Rad).
2. Towbin transfer buffer: 25 mM Tris, 192 mM glycine, and 20% methanol.
3. Nitrocellulose membrane.
4. Blotting paper.
5. 10× Tris-buffered saline (TBS): 1.5 M sodium chloride (NaCl) and 0.1 M Tris pH 7.4.
6. 1× TBS-T: 1× TBS, 0.05% Tween 20.
7. 5% nonfat dry milk dissolved in 1× TBS-T.
8. 2.5% bovine serum albumin (BSA) dissolved in 1× TBS-T.
9. Enhanced chemiluminescence (ECL) substrate, such as Western Lightning® Plus-ECL (PerkinElmer).
10. 7 M guanidine hydrochloride.
11. CYLD rabbit monoclonal antibody D1A10 (Cell Signaling Technology).
12. β-ACTIN mouse monoclonal antibody 8H10D10 (Cell Signaling Technology).
13. Anti-rabbit IgG conjugated to horseradish peroxidase (HRP) (Jackson ImmunoResearch Laboratories).
14. Anti-mouse IgG conjugated to horseradish peroxidase (HRP) (Santa Cruz Biotechnology).
15. Laboratory rocker.
16. Orbital shaker.

3 Methods

3.1 Stimulation

1. Remove the BMDMs from 10 cm non-tissue culture treated plates using 1 mL 0.25% trypsin containing 2.21 mM EDTA per plate. Incubate at 37 °C with 5% CO_2 for approximately 5 min, or until the cells are lifted from the plate (*see* **Note 2**).
2. Deactivate the trypsin by adding 10 mL of RPMI complete media to the plate and gently pipetting to dislodge cells from the plate. Transfer to a conical tube and centrifuging at 250 × *g* for 5 min. Discard supernatant.
3. Wash the cells once by resuspending in 10 mL of 1× PBS, centrifuging at 250 × *g* for 5 min, and removing the supernatant.
4. Resuspend the cells in RPMI complete containing 30% L929-CM and count the cells.
5. Plate the BMDMs in 24-well tissue culture treated plates at a density of 0.25 × 10^6 per well in 0.5 mL. Allow the cells to attach to the bottom of the well by culturing for an additional 24 h.
6. To conduct a time-course study on the effects of LPS on the cleavage of CYLD p107 to CYLD p25, stimulate the BMDMs with a fixed dose of LPS from *E. coli* 0111:B4 (e.g., 100 ng/mL) for time points ranging from 0 to 8 h.
7. To conduct a dose-response study, stimulate the BMDMs with *E. coli* LPS for a fixed period of time (e.g., 4 h), with LPS doses ranging from 0 to 200 ng/mL.

3.2 Cell Lysis

1. After stimulation, wash the adherent cells with 0.5 mL of 1× PBS per well.
2. To lyse the cells, add 50 μL of ice cold cell lysis buffer to each well and incubate on ice for 10–15 min, with occasional shaking.
3. Transfer the lysate to 1.5 mL microcentrifuge tube and centrifuge at 10,000 × *g* for 10 min at 4 °C. The supernatant will contain the proteins of interest.
4. Determine the protein concentration of the prepared lysate using Pierce BCA Protein Assay Kit.

3.3 Preparation and Running of SDS-PAGE Gels

1. Clean and assemble the Mini-Protean® system.
2. Prepare 10 mL of 10% resolving gel (*see* **Note 3**). Immediately pour into the space between the glass plates. Leave 1 cm of space below the bottom of the comb for the stacking gel. Overlay the resolving gel with a thin layer of distilled water or water-saturated 1-butanol. Leave at room temperature for 30 min to polymerize.

3. Pour and wash off the water or 1-butanol from the resolving gel.
4. Prepare 5 mL of 5% stacking gel. Mix and immediately pour on top of the polymerized resolving gel. Insert comb, being careful to avoid air bubbles. Allow the gel to polymerize at room temperature for 30 min.
5. Meanwhile, aliquot 50 μg of protein from each lysate into a microcentrifuge tube and if needed, bring the total volume up to 30 μL using lysis buffer.
6. Add 7.5 μL of the 5× loading buffer so that the final concentration is 1×. Heat samples for 5 min at 95 °C to denature the proteins. Centrifuge to bring samples to the bottom of the tube.
7. Assemble the Mini-Protean® system and fill chambers with 1× Tris/Glycine/SDS running buffer. Carefully load the prestained protein ladder and denatured lysates into individual wells.
8. Attach the apparatus to a power supply and run at 120 V until the bromophenol blue dye reaches the bottom of the glass plates (*see* **Note 4**).

3.4 Transfer and Immunoblotting

1. Soak the blotting papers, nitrocellulose membrane, and fiber pads in Towbin transfer buffer.
2. Arrange in the transfer cassette in the following order, while soaking in transfer buffer:

   Fiber pad.

   Blotting paper.

   Nitrocellulose membrane (cut the upper right side of the membrane to ensure the correct orientation of the membrane throughout the immunoblotting process).

   Gel (protein side up, so the samples are oriented from the left to the right).

   Blotting paper.

   Fiber pad.
3. Submerge the cassette and its contents in the transfer tank filled with transfer buffer. Ensure that the membrane is in between the gel and the anode. Place the transfer tank on ice and run at 100 V for 1 h.
4. Disassemble the apparatus and the transfer cassette.
5. Rinse the membrane once in 1× TBS-T.
6. Incubate the membrane in 5% nonfat dry milk in 1× TBS-T at room temperature for 2 h on a laboratory rocker, or at 4 °C overnight on a laboratory rocker.

7. Dilute the anti-CYLD (D1A10) antibody 1:1000 in 2.5% BSA in 1× TBS-T (*see* **Note 5**).
8. Incubate the membrane in this antibody dilution at room temperature for 2 h on a rocker.
9. Vigorously wash the membrane twice in a large volume of 1× TBS-T for 15 min each on an orbital shaker at room temperature.
10. Dilute the anti-rabbit IgG HRP secondary antibody 1:15,000 in 2.5% milk in 1× TBS-T (*see* **Note 6**).
11. Incubate the membrane in this antibody dilution at room temperature for 1 h on a rocker.
12. Vigorously wash the membrane twice in a large volume of 1× TBS-T for 15 min each.
13. Mix the ECL reagents in a ratio of 1:1 and submerge the membrane in this mixture for 1 min.
14. Immediately expose the membrane to X-ray film to visualize protein expression. An increase in the intensity of a CYLD-reactive band of approximately 25 kDa following stimulation is indicative of CASPASE 8 enzymatic activity and suppression of necroptosis (*see* **Note 7**). A representative blot is shown in Fig. 1.
15. To control for equivalent loading of wells, reprobe the membrane for expression of β-ACTIN. Strip the membrane by submersion in 7 M guanidine hydrochloride and agitation on an orbital shaker for 10 min (*see* **Note 8**).

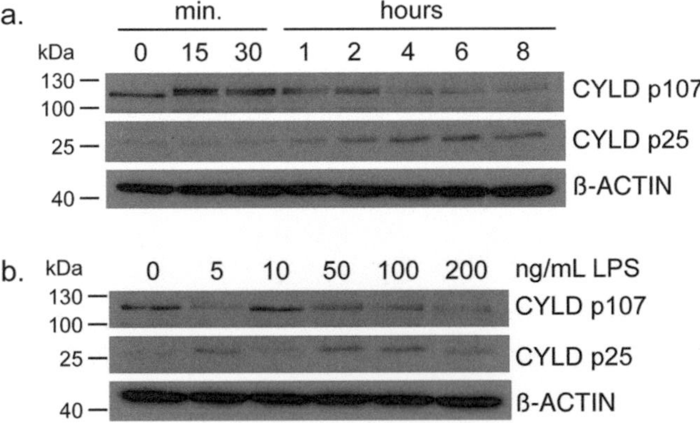

Fig. 1 BMDMs were stimulated with 100 ng/mL of LPS for the indicated times (**a**) or with the indicated concentrations of LPS for 4 h (**b**). Lysates were subject to SDS-PAGE and western blot analysis using CYLD and β-ACTIN antibodies. The experiments shown are representative of at least three independent experiments

16. Wash the membrane twice in a large volume of 1× TBS-T for 15 min each.
17. Repeat immunoblotting procedure using anti-β-ACTIN (1:1000 dilution) as the primary antibody and anti-mouse IgG HRP (1:10,000 dilution) as the secondary antibody.

4 Notes

1. It is best to add the protease cocktail inhibitors to the lysis buffer immediately before use.
2. Since BMDMs are difficult to trypsinize, it may take longer than 5 min of incubation with trypsin-EDTA to lift the BMDMs from the plate.
3. Unpolymerized acrylamide is a neurotoxin and suspected carcinogen, and so precautions must be taken when handling it.
4. If the BioRad Mini-Protean® system is being used, at 120 V it should take 90 min for the bromophenol blue dye to reach the bottom of the glass plates.
5. The primary antibody can be saved and reused three times if 0.05% sodium azide is added to the antibody dilution and it is stored at 4 °C. Note that sodium azide is very toxic, and so extreme caution should be taken when handling it.
6. Sodium azide will inactivate HRP, and so it should not be added to the secondary antibody dilution.
7. Dependency of the CYLD cleavage on CASPASE 8 activity can be shown by including samples in which the cells are pretreated with the pancaspase inhibitor z-VAD-FMK prior to stimulation, which will inhibit the appearance of the 25 kDa fragment. The more specific inhibitor of CASPASE 8, z-IETD-FMK, can be similarly used.
8. Guanidine hydrochloride can be reused up to three times, or until the solution becomes yellow. It is an irritant when it comes in contact with skin, and so care should be taken when handling it.

Acknowledgments

This work was supported by grants AI052417 and DK072201 from the NIH, and a Senior Research Award from the Crohn's and Colitis Foundation of America (CCFA).

References

1. Hitomi J, Christofferson DE, Ng A, Yao J, Degterev A, Xavier RJ, Yuan J (2008) Identification of a molecular signaling network that regulates a cellular necrotic cell death pathway. Cell 135(7):1311–1323. https://doi.org/10.1016/j.cell.2008.10.044
2. O'Donnell MA, Perez-Jimenez E, Oberst A, Ng A, Massoumi R, Xavier R, Green DR, Ting AT (2011) Caspase 8 inhibits programmed necrosis by processing CYLD. Nat Cell Biol 13(12):1437–1442. https://doi.org/10.1038/ncb2362
3. Vanlangenakker N, Vanden Berghe T, Bogaert P, Laukens B, Zobel K, Deshayes K, Vucic D, Fulda S, Vandenabeele P, Bertrand MJ (2011) cIAP1 and TAK1 protect cells from TNF-induced necrosis by preventing RIP1/RIP3-dependent reactive oxygen species production. Cell Death Differ 18(4):656–665. https://doi.org/10.1038/cdd.2010.138
4. Legarda D, Justus SJ, Ang RL, Rikhi N, Li W, Moran TM, Zhang J, Mizoguchi E, Zelic M, Kelliher MA, Blander JM, Ting AT (2016) CYLD proteolysis protects macrophages from TNF-mediated auto-necroptosis induced by LPS and licensed by type I IFN. Cell Rep 15(11):2449–2461. https://doi.org/10.1016/j.celrep.2016.05.032

INDEX

A

Acetonitrile...118, 163, 165–167
Acute kidney injury (AKI)...135–143, 149, 153
Anaesthetics...136, 139, 140, 143, 147, 148
Analgesic...136, 138, 140, 143, 147
Annexin-V...4, 37, 40–45, 47, 48, 73, 79, 132
Aorta...141, 148
AP 20187...72, 73, 76, 78, 81
Apoptosis...v, 1, 3, 4, 6, 15, 17, 19, 20, 23–24, 35–50, 53, 90, 94, 101, 102, 112, 116, 117, 126, 142, 171, 172
Autophosphorylation...71, 153

B

Bicinchoninic acid assay (BCA)...88, 104, 112, 182, 184
Biotin...168
Bismaleimidohexane (BMH)...73, 77, 80
Blue-native PAGE...55, 58, 59
Bone marrow derived macrophages (BMDM)...63–69, 102, 106, 111–112, 114–117, 126, 127, 129, 131, 132, 181–187
Bovine serum albumin (BSA)...21, 77, 88, 103, 105, 113, 155, 157, 183, 186
Buprenorphine–HCl...136, 140, 143, 147, 148

C

Caspase 1...47
Caspase 4...4
Caspase 5...4
Caspase 8...16, 19, 23, 36, 53–55, 57, 60, 63, 71, 81, 94, 102, 106, 126, 154, 172, 181–187
Caspase 11...4
Caspase inhibitors
  IDN-6556/Emricasan...20
  qVD-OPh...20
  zVAD-fmk...3, 16, 60, 102, 106, 116, 126
Cavitron (2-hydroxypropyl)-β-cyclodextrin...112, 122
Cell lines
  Human HT29 adenocarcinoma cells...21
  Human Jurkat T cells...21
  Human U937 monocytic cells...21
  Mouse L929 fibrosarcoma cells...21
  Mouse mesangial MES 13...146
  Mouse NIH-3T3 fibroblast cells...76, 77, 80
  Mouse RAW264.7 macrophages...21
CellTiter 96® Aqueous One Solution Cell Proliferation Assay...12, 13, 15, 17, 128, 130
CellTiter-Glo® Luminescent Cell Viability Assay...21, 65, 68, 69
Cellular inhibitors of apoptosis (cIAPs)...3, 5, 53, 54, 171, 172
Chemical library...4, 19–32
Chloroacetic acid...163, 165, 167
Chromatography...161–168
Cisplatin...141–142
Concanavalin A (Con A)...147
Coverslips...155, 157, 158
Cycloheximide (CHX)...20, 30, 94, 95, 97, 102, 103, 106, 112, 117, 132
Cylindromatosis (CYLD)...3, 5, 11, 181–187

D

Damage-associated molecular patterns (DAMP)...v, 2, 4, 35, 125
Death inducing signaling complex (DISC)...101–106
Death receptors (DR)...2, 3, 101, 125
Dehydrant...154, 156, 157, 159
Dithiothreitol (DTT)...21, 87, 88, 91, 163, 173, 175, 182
Doxycycline (DOX)...73, 76, 78, 79, 81

E

Electrospray ionization (ESI)...167
Embedding...154, 156, 159
Embryos...75, 102, 154, 157, 159
Endothelial cells...147–148, 154
Ethanol...64, 65, 103, 111, 114, 130, 154–157, 159

F

FAS...101, 102
Fas-associated protein with death domain (FADD)...4, 12, 14, 22, 23, 30, 54, 71, 93, 98, 101–106, 172
FKBP binding domains...74
FLAG®...163
FLAG-tagged human TNF (FLAG-TNF)...172, 175
Focal segmental glomerulosclerosis (FSGS)...148–149
Forceps...111, 136–138, 140, 147, 156

Adrian T. Ting (ed.), *Programmed Necrosis: Methods and Protocols*, Methods in Molecular Biology, vol. 1857,
https://doi.org/10.1007/978-1-4939-8754-2, © Springer Science+Business Media, LLC, part of Springer Nature 2018

Formalin ... 154, 155, 159
Formic acid ... 163, 166, 167

G

Glomerular Injury ... 145–149
Glomeruli ... 146, 147, 149
Glomerulonephritis ... 145
Glomerulopathy ... 145
Glomerulosclerosis ... 148, 149
Glomerulus ... 148
Glycerol ... 21, 38, 55, 87, 95, 103, 142, 143, 162, 163, 172, 175, 182

H

Hematoxylin ... 155, 157, 158

I

Immunoblotting ... 38, 39, 49, 86–87, 90, 91, 94–95, 103, 105, 174, 176, 183, 185–187
Immunohistochemistry ... 5, 153–159
Immunoprecipitation (IP) ... 4, 21, 28, 29, 74, 81, 94–98, 102–104, 106, 112, 118, 123, 154, 162, 164–167, 173–177
Influenza A ... 93–97
Inhibitor of apoptosis (IAP) ... 19
Inhibitor of kappaB kinase (IKK) ... 177
Interferon-beta (IFNβ) ... 20
Interferon-gamma (IFNγ) ... 20
Iodoacetamide ... 167
Ischemia reperfusion injury ... 135–139
Isoflurane ... 136, 140, 143, 147, 148

K

Keratin ... 162, 167
Kidney ... 5, 98, 135–143, 145–149, 153

L

Laemmli ... 26, 28, 29, 31, 32, 87, 88, 91, 173–175
Lambda Protein Phosphatase (λPPase) ... 173, 175–177
Linear UBiquitin chain Assembly Complex (LUBAC) ... 53, 171
Lipopolysaccharide (LPS) ... 20, 23–31, 63, 64, 66–68, 106, 126, 128–132, 146, 147, 181, 182, 184, 186
Liquid chromatography-tandem mass spectrometry (LC-MS/MS) ... 118, 122, 161
LoBind ... 163–165, 168
Lysyl endopeptidase/Lys-C ... 163

M

Macrophages ... 4, 21, 23, 28–30, 63–69, 106, 111–112, 114–115, 126, 127, 129, 181–187
Mass spectrometry ... 161–168
Mesangial cells (MC) ... 145–148
Mesangiolysis ... 145–147, 149
Microaneurysm clamp ... 136, 137, 147
Micro-cannula ... 146–148
Microtome ... 155, 156, 159
Mitogen-activated protein kinase (MAPK) ... 53, 171
Mixed lineage kinase domain-like (MLKL) ... 2, 11, 19, 35, 54, 71, 72, 76, 78, 85, 93, 102, 122, 125, 153
Modified tandem affinity purification (Mo-TAP) ... 161–168
Mouse embryonic fibroblasts (MEF) ... 21, 75, 88, 91, 94, 102, 132, 173, 174

N

Necroptosis ... v, 1–6, 11–13, 15, 17, 19–32, 35–50, 53–60, 63–69, 71–81, 85, 90, 93, 94, 101–106, 109–123, 125–133, 135, 142, 149, 153–159, 171, 172, 181, 186
Necrosome ... 4, 21, 25–30, 54, 81, 85, 93–97, 102, 122
Necrosulfonamide (NSA) ... 37, 39–42, 44–45, 47, 48
NF-kappa-B essential modulator (NEMO) ... 172
Nonidet P-40 (NP-40) ... 74, 103, 172–174
Nuclear factor-kappaB (NF-kB) ... 53, 112, 117

P

Paraffin ... 154, 156, 158, 159
Peptide mass fingerprints (PMFs) ... 162, 167
Pharmacokinetic (PK) ... 110–113, 117–122, 136, 138, 140, 143, 147
Phosphatidylserine ... 41, 47, 48
Phosphorylation ... 3, 5, 36, 54, 60, 72, 73, 78
Podocyte ... 148–149
Polyubiquitin ... 53, 181
PreScission ... 162–164
Programmed necrosis ... v, 1–2, 11, 85–91
Propidium iodide (PI) ... 4, 37, 39–45, 47, 48, 55, 56, 58, 64–67, 69, 73, 79, 132
Proteolysis ... 181–187
Pseudokinase ... 2, 72, 125
Pyroptosis ... 2, 47

R

Receptor-interacting protein kinase 1 (RIPK1) ... 2, 12, 19, 36, 53, 63, 71, 85, 93, 102, 109, 125, 171, 181
Receptor-interacting protein kinase 3 (RIPK3) ... 2, 11, 19, 35, 54, 63, 71, 72, 76, 85, 93, 96, 102, 125, 153
Regulated cell death (RCD) ... v, 1, 2, 4–6, 35, 148
Retractors ... 136, 137, 139, 140
Rhabdomyolysis ... 142–143
RIP homotypic interaction motifs (RHIM) ... 2, 3, 54, 71, 93
RIPA lysis buffer ... 38, 74
RIPK1 inhibitors
  GSK'481 ... 110, 111, 123
  GSK'963, 110, 111, 123
  Necrostatin-1 (Nec-1), 125
  Necrostatin-1S (nec-1s), 37

RIPK3 inhibitors
 dabrafenib 72
 GSK'843 72
 GSK'872 ....23, 24, 37, 39, 40, 42, 44–45, 72, 73
 GW440139B (GW'39B) 72, 73
RNA interference (RNAi) 4, 11–18

S

Scissor 64, 111, 113, 115, 123, 136–138, 140, 147, 148
Second mitochondria-derived activator of caspases (SMAC) 3, 5, 20, 37, 40, 53, 54, 56, 60, 85, 87, 88, 90, 102, 114, 132
SHANK associated RH domain interactor (SHARPIN) 171, 172
Short hairpin RNAs (shRNA) 11, 12
SiRNA library 12–14
SMAC mimetics
 AZD5582 dihydrochloride 37
 birinapant 53, 54
 compound A 30, 54, 60
 SM164 ....20, 23, 30
Small interfering RNAs (SiRNA) 11–17
Sodium dodecyl sulfate-polyacrylamide gel electrophoresis (SDS-PAGE) 26, 32, 38, 39, 46, 55, 57, 74, 75, 77, 86, 88, 95, 97, 105–106, 112, 117, 182–186
Staurosporine 12, 15, 17
Strep 73, 162, 164, 165
Strep-Tactin 163, 165, 166, 168
Suture 136, 137, 139, 140, 147, 148
Syto 24 Green 73, 79
Sytox Green 73, 79, 81

T

TAK1 inhibitor
 5z-7-oxozeaenol 20, 23, 30
TNFR-associated factor 2 (TRAF2) 53
TNF receptor associated death domain (TRADD) 53, 54, 101, 172, 174
TNF-related apoptosis inducing ligand (TRAIL) 20, 125
Toll-like receptors (TLR) 20, 125, 126
Transforming growth factor beta-activated kinase 1 (TAK1) 20, 172, 177
Trifluoroacetic acid (TFA) 163, 165, 166
Triton 21, 26, 38, 94, 167, 182
Tumor necrosis factor (TNF) 12, 63, 85, 128, 162, 171, 181
Tumor necrosis factor receptor 1 (TNFR1) 3, 53, 79, 102, 125, 161–168, 171–177, 181
Tumor necrosis factor receptor 2 (TNFR2) 12, 14, 16, 175

U

Ubiquitin 53, 171, 176, 177
Ubiquitination 3, 5, 117, 167, 181
Ubiquitin carboxyl-terminal hydrolase 2 (USP2) 173, 175–177
Unilateral ureteric obstruction (UUO) 139–141
Urea 136, 163, 165, 166, 168

V

Viability assay 17, 21, 23, 30, 64, 65, 67–69, 110, 116, 127–130, 132

X

Xylene 154–157

Z

ZVAD-fmk (z-Val-Ala-Asp-(OMe)-Fluoromethyl Ketone) 3, 12, 16, 57, 58, 60, 64, 73, 87, 102, 103, 106, 113, 116, 126

MIX
Papier aus verantwortungsvollen Quellen
Paper from responsible sources
FSC® C105338

If you have any concerns about our products, you can contact us on
ProductSafety@springernature.com

In case Publisher is established outside the EU, the EU authorized representative is:
Springer Nature Customer Service Center GmbH
Europaplatz 3, 69115 Heidelberg, Germany

Printed by Libri Plureos GmbH
in Hamburg, Germany